The Philosophy of Human Evolution

This book provides a unique discussion of human evolution from a philosophical viewpoint, looking at the facts and interpretations since Charles Darwin's *The Descent of Man*. Michael Ruse explores such topics as the nature of scientific theories, the relationships between culture and biology, the problem of progress, and the extent to which evolutionary issues pose problems for religious beliefs. He identifies these issues, highlighting the problems for morality in a world governed by natural selection. By taking a philosophical viewpoint, the full ethical and moral dimensions of human evolution are examined. This book engages the reader in a thorough discussion of the issues, appealing to students in philosophy, biology, and anthropology.

MICHAEL RUSE is the Lucyle T. Werkmeister Professor of Philosophy and the Director of the Program in the History and Philosophy of Science at Florida State University. His previous publications include *Can a Darwinian Be a Christian? The Relationship between Science and Religion* (Cambridge, 2004), *Darwinism and Its Discontents* (Cambridge, 2008), and *Science and Spirituality: Making Room for Faith in an Age of Science* (Cambridge, 2010).

**Cambridge Introductions to
Philosophy and Biology**

General editor
Michael Ruse, Florida State University
Associate editor
Denis Walsh, University of Toronto

Other titles in the series

Derek Turner, *Paleontology: A Philosophical Introduction*
R. Paul Thompson, *Agro-Technology: A Philosophical Introduction*
Michael Ruse, *The Philosophy of Human Evolution*

The Philosophy of Human Evolution

MICHAEL RUSE

Florida State University

CAMBRIDGE
UNIVERSITY PRESS

CAMBRIDGE UNIVERSITY PRESS
Cambridge, New York, Melbourne, Madrid, Cape Town,
Singapore, São Paulo, Delhi, Tokyo, Mexico City

Cambridge University Press
The Edinburgh Building, Cambridge CB2 8RU, UK

Published in the United States of America by Cambridge University Press, New York

www.cambridge.org
Information on this title: www.cambridge.org/9780521117937

© Michael Ruse 2012

First published 2012

Printed in the United Kingdom at the University Press, Cambridge

A catalogue record for this publication is available from the British Library

Library of Congress Cataloguing in Publication data
Ruse, Michael.
  The philosophy of human evolution / Michael Ruse.
    p.   cm. – (Cambridge introductions to philosophy and biology)
  Includes bibliographical references and index.
  ISBN 978-0-521-11793-7 (hardback) – ISBN 978-0-521-13372-2 (paperback)
  1. Human evolution – Philosophy.   2. Human evolution – Social aspects.   I. Title.
  GN281.R84 2012
  599.93′8–dc23
  2011039357

ISBN 978-0-521-11793-7 Hardback
ISBN 978-0-521-13372-2 Paperback

For Francisco Ayala

Contents

Figures

Acknowledgements

No book is written in isolation, certainly not this one. As you will see on every page, it was Charles Darwin and his wonderful theory of evolution through natural selection that made possible all interesting inquiry about human origins and their implications. More immediately, for detailed knowledge about human evolution I am much in the debt of my colleague, the paleoanthropologist Dean Falk. Her wisdom and friendship are very much appreciated. As always, my thinking has been shaped in many ways by the interaction with three people: the philosopher David Hull, the historian and philosopher Robert J. Richards, and the biologist Edward O. Wilson. The paleontologist Daniel McShea has long stimulated my interest in the question of progress in evolution, and I owe much to Randolph Nesse for his pioneering thinking about evolutionary medicine.

Hilary Gaskin at Cambridge University Press was supportive as always, and the co-editor of the series to which this book belongs, Denis Walsh, gave the manuscript a thoughtful reading. I am tremendously in debt to my wife Lizzie for everything. Finally, I want say what pleasure my dedication gives to me. I have known Francisco J. Ayala for over thirty years, admired his work, and enjoyed his friendship. Above all, I respect the time and effort he has given to the well-being of science, whether it be on the witness stand critiquing Creationism, taking leadership roles in leading organizations and academies, or welcoming people from the world over to his laboratory. I doubt whether he will agree with everything I have said in this book, but I can say that without the work of people like him such a book could not have been written.

1   Evolutionary biology

What is "evolution"? If you were asking in the middle of the eighteenth century, the answer would be something to do with individual development. But from about the middle of the nineteenth century the term has meant a gradual process of law-bound development that brought about the world in which we live, especially the world of organisms, animals, and plants (Richards 1992). We usually add that we mean a natural process, that is something where supernatural forces like God play no direct role. Often we mean common ancestry, a "tree of life," although not every evolutionist has believed in such a tree. Today we think more in terms of bushes or sometimes perhaps a very odd tree like the banyan tree, where there are links all over the place.

Early years

The ancient Greeks did not believe in evolution (Sedley 2008). In a way, this seems strange, because the greatest philosophers – Plato (428–348 BCE) and Aristotle (384–322 BCE), in particular – saw this world of ours as one of change and motion. But when one person, the Sicilian pre-Socratic philosopher Empedocles (490–430 BCE), did suggest something along developmental lines, Aristotle criticized him severely. The reason is simple. The Greeks saw the world as designed, as put together for purposes, for ends (Ruse 2003). Organisms are the things in the world that more than any other exhibit what Aristotle called "final causes." If you want to understand something like the hand or the eye, then you must ask not merely what the forces making them were, but what is the reason for their existence. Hands and eyes are complex things, and could not have come about through mere chance – through the action of blind, undirected law – but require an intelligence of some kind to make them. Hence, evolution – the epitome of blind chance – is impossible.

The coming of Christianity reinforced this mind-set. The early Christians were not at first sure that they needed to accept the Hebrew Scriptures – after all, it was the Jews who rejected Jesus. But, particularly under the influence of St. Augustine (354–430 CE), they realized that it is only through the Jewish narratives that sense can be made of Jesus and his fate. Why did he have to die on the cross? For our sins. Why are we sinful? Because of the acts of Adam and Eve, as told to us in Genesis. Augustine particularly cautioned that one should not necessarily interpret Scripture literally, especially if it conflicts with modern science. But then, and for many centuries later, no one had reason to doubt the authenticity of the Genesis creation story, which puts all down to miracle, not that long ago.

It was only at the beginning of the Enlightenment, the flowering of science and philosophy that is generally dated from the beginning of the eighteenth century, that people first started to speculate seriously in the direction of developmental origins (Ruse 1996, 2005a). It is true that increasingly there were empirical discoveries that today we would think evidence of evolution – fossils and strange transitional creatures, particularly – but then and for many years afterwards the chief attraction of evolution was ideological. The Enlightenment was the time when many people first began seriously to adopt the philosophy of Progress – the belief that through unaided human talent and effort the human condition, science, medicine, teaching, culture, and more, can be improved. The Enlightenment was the time when many people began to reject the older philosophy of Providence – the belief that we humans unaided can do nothing without God's help and to think otherwise is presumptuous and doomed to disaster. Evolution – now, without qualification, meaning the evolution of organisms – was caught up in this debate about Progress and Providence, with supporters of the former arguing that as we see Progress in the cultural world, so likewise we see progress in the biological world, going all of the way from blobs to humans, from monads to man as the saying went. (Progress with a capital "P" will refer to cultural notions of upward change; progress with a small "p" to biological notions of such change.)

Note that although evolution was taken to be something against the core beliefs of Christianity, it was not simply a matter of denying the literal words of the Bible. Augustine had prepared the way for people to do this. It was more one of going against what was seen to be the essential relationship between God and his creatures. Not that the early evolutionists were atheists, or even agnostics. Rather, they believed in a God who works through unbroken law – a God of "deism" as opposed to the interventionist God of "theism" – and of course for such a God, the law-bound process of evolution is confirmation of his greatness rather than refutation. Note also that the P/progressionists were as convinced of the central importance of humankind as were the Providentialists. The late eighteenth-century evolutionist Erasmus Darwin (1731–1802), grandfather of Charles Darwin, shows this clearly. He was much given to expressing his ideas in verse.

> Organic Life beneath the shoreless waves
> Was born and nurs'd in Ocean's pearly caves;
> First forms minute, unseen by spheric glass,
> Move on the mud, or pierce the watery mass;
> These, as successive generations bloom,
> New powers acquire, and larger limbs assume;
> Whence countless groups of vegetation spring,
> And breathing realms of fin, and feet, and wing.
>
> Thus the tall Oak, the giant of the wood,
> Which bears Britannia's thunders on the flood;
> The Whale, unmeasured monster of the main,
> The lordly Lion, monarch of the plain,
> The Eagle soaring in the realms of air,
> Whose eye undazzled drinks the solar glare,
> Imperious man, who rules the bestial crowd,
> Of language, reason, and reflection proud,
> With brow erect who scorns this earthy sod,
> And styles himself the image of his God;
> Arose from rudiments of form and sense,
> An embryon point, or microscopic ens!
>
> (E. Darwin 1803, I, lines 295–314)

He was also unambiguous about the way in which he tied his biology into his philosophy. This idea of organic progressive evolution "is analogous to the improving excellence observable in every part of the creation ... such as the progressive increase of the wisdom and happiness of its inhabitants" (E. Darwin 1794–96, 509).

The opening of the nineteenth century saw many other people embracing evolutionary ideas. ("Transmutation" was a popular term for the idea.) Probably the most famous was the French biologist Jean Baptiste de Lamarck (1744–1829), who gave his name to the process of change that supposes that acquired characteristics (like the blacksmith's strong arm) can be inherited directly. But enthusiasm generally outran evidence, and virtually no one even tried to speak to the problem of final causes. No one really grappled with the Greeks' problem, how can blind law bring on complex functioning? It was not until 1859 that this all changed, when the British naturalist Charles Robert Darwin (1809–82) published his great work *On the Origin of Species by Means of Natural Selection, or the Preservation of Favoured Races in the Struggle for Life*. Now finally the world was presented with a full-bodied attempt to explain origins, of organisms living and dead, in purely natural terms, at the same time speaking to the Aristotelian worries about the functioning of organisms. More precisely, Darwin tried to make reasonable or commonsensical the very fact of evolution – all organisms have a common descent by natural processes from one or just a few original simple forms – and at the same time provide a mechanism of change that speaks to final cause, this mechanism being something Darwin called "natural selection."

Charles Robert Darwin

In some respects Darwin does not seem a very likely candidate for the role of the "father of evolution," as he is often called (Browne 1995, 2002). Born into a rich, upper-middle-class English family – his maternal grandfather was Josiah Wedgwood, responsible for the modernization of the British pottery trade – he had conventional schooling, aiming first to be a physician (like his father) and then when that failed to be a clergyman in the state-established Church of England. Things changed when, after university (Cambridge), Darwin spent five years on board the British warship HMS *Beagle*, as she mapped the coast of South America. As ship's naturalist, he made massive collections of specimens, and developed rapidly into a full-time scientist, primarily in the early years as a full-time geologist, but then more and more as a biologist. We know that he became an evolutionist and discovered the mechanism for which he is famous, natural selection, in the late 1830s; but for reasons that are still not entirely clear he did not publish for another twenty years. Instead, he married his first cousin Emma Wedgwood, moved to the country outside London, raised a large family, and spent much of his time battling unknown, but very draining, bodily ailments. Evolution was put to one side as Darwin engaged in a massive study of barnacle taxonomy, and it was only when that was finished (in the 1850s) that Darwin turned again to evolution. Famously, he was finally spurred into print when a young naturalist in the Malay Archipelago, Alfred Russel Wallace (1823–1913), sent to Darwin of all people a short essay which showed that quite independently he had hit upon the same mechanisms as the older man. In fifteen months, Darwin wrote the *Origin of Species*, which was published late in the year 1859.

In his *Autobiography*, written late in life, Darwin spoke of the *Origin* as containing "one long argument" (C. Darwin 1958, 140). But what was this argument? Actually it came in several (at least three) parts. In a letter written a year or two after the *Origin* was first published, Darwin made explicit mention of his strategy.

> In fact the belief in natural selection must at present be grounded entirely on general considerations. (1) on its being a vera causa, from the struggle for existence; & the certain geological fact that species do somehow change (2) from the analogy of change under domestication by man's selection. (3) & chiefly from this view connecting under an intelligible point of view a host of facts. (Letter to George Bentham, May 22, 1863; C. Darwin 1985–, XI, 433)

Let's start there. The *Origin* went through six, increasingly revised, editions, although most scholars today prefer the first (1859), untouched version. As you open the *Origin* you find that in fact it is the analogy with the domestic world that comes first. Probably reflecting his personal route to discovery, Darwin pointed out at length that breeders have taken organisms like pigeons and dogs, like cattle and sheep, not to mention vegetables and other plants, and simply transformed them, creating many different forms and varieties. And that this has all been done by means of taking those with features one desires – shaggier coats, prettier feathers, fiercer fighting natures – and breeding from and only from these specimens.

> The great power of this principle of selection is not hypothetical. It is certain that several of our eminent breeders have, even within a single lifetime, modified to a large extent some breeds of cattle and sheep.

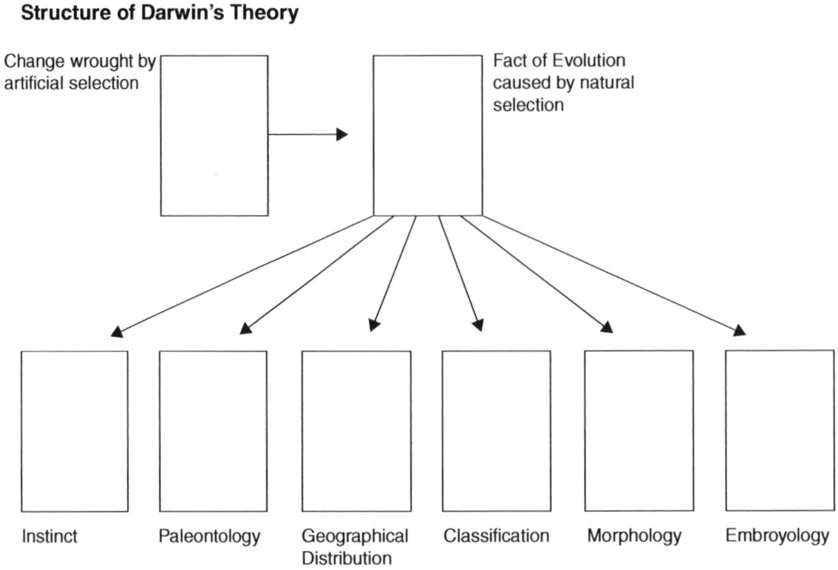

Structure of Darwin's Theory

Change wrought by artificial selection

Fact of Evolution caused by natural selection

Instinct   Paleontology   Geographical Distribution   Classification   Morphology   Embroyology

Figure 1.1 The structure of the *Origin*. Note first the analogy between the world of the breeder and the world of nature, and how Darwin then uses the central mechanism of evolution through natural selection to explain in different areas of biology and conversely uses these explanations as support for his central mechanism.

> In order fully to realize what they have done, it is almost necessary to read several of the many treatises devoted to this subject, and to inspect the animals. Breeders habitually speak of an animal's organisation as something quite plastic, which they can model almost as they please. If I had space I could quote numerous passages to this effect from highly competent authorities. (C. Darwin 1859, 31)

Next, picking up on the first point he mentioned in the letter to Bentham (a nephew, incidentally, of the philosopher Jeremy Bentham), Darwin introduced his main mechanism of natural selection. He did this in two parts. First, arguing from the tendency of organisms to multiply in number, he agreed with the political philosopher Thomas Robert Malthus that because space and food will always be limited, there are going to be inevitable "struggles for existence."

> A struggle for existence inevitably follows from the high rate at which all organic beings tend to increase. Every being, which during its natural lifetime produces several eggs or seeds, must suffer destruction during

some period of its life, and during some season or occasional year, otherwise, on the principle of geometrical increase, its numbers would quickly become so inordinately great that no country could support the product. Hence, as more individuals are produced than can possibly survive, there must in every case be a struggle for existence, either one individual with another of the same species, or with the individuals of distinct species, or with the physical conditions of life. It is the doctrine of Malthus applied with manifold force to the whole animal and vegetable kingdoms; for in this case there can be no artificial increase of food, and no prudential restraint from marriage. Although some species may be now increasing, more or less rapidly, in numbers, all cannot do so, for the world would not hold them. (C. Darwin 1859, 63–64)

Darwin then drew on the fact that whenever you have a population of organisms, that is a group all in the same species, you find nevertheless that there are differences between them and that every now and then something new seems to pop up into being. This led Darwin to speculate that in the struggle some types or forms are likely to prove more success-ful than others, simply because these types or forms will help their pos-sessors against others. Given enough time, these types will spread through the group and eventually there will be full-blooded change.

Can the principle of selection, which we have seen is so potent in the hands of man, apply in nature? I think we shall see that it can act most effectually. Let it be borne in mind in what an endless number of strange peculiarities our domestic productions, and, in a lesser degree, those under nature, vary; and how strong the hereditary tendency is. Under domestication, it may be truly said that the whole organisation becomes in some degree plastic. Let it be borne in mind how infinitely complex and close-fitting are the mutual relations of all organic beings to each other and to their physical conditions of life. Can it, then, be thought improbable, seeing that variations useful to man have undoubtedly occurred, that other variations useful in some way to each being in the great and complex battle of life, should sometimes occur in the course of thousands of generations? If such do occur, can we doubt (remembering that many more individuals are born than can possibly survive) that individuals having any advantage, however slight, over others, would have the best chance of surviving and of procreating their kind? On the other hand, we may feel sure that any variation in the least degree injurious would be rigidly destroyed. This preservation of favourable

variations and the rejection of injurious variations, I call Natural
Selection. (80–81)

Note a point made above but sufficiently important to be worth making a
second time. Darwin's natural selection does not just bring about change; it
brings about change of a particular kind. Organisms are adapted – they have
adaptations, features that aid them in the struggle to survive and reproduce.
Hands, eyes, teeth, penises, vaginas, leaves, flowers, seeds, and more – these
are things that are "as if" designed, that is to say that they are put together
in order to help their possessors. And it is Darwin's claim that this all comes
about naturally. There is no need to invoke God or any other force making
for the design-like nature of the organic world. Darwin does not deny final
causes, not at all. He simply wants to give them a natural beginning.

At this point, Darwin introduced a secondary form of selection, sexual
selection. Just as natural selection is modeled on the features in the world
of the breeder that might be expected to have analogous roles in survival
and reproduction against competitors and elements in the natural world –
thicker coats, better egg-laying abilities – so sexual selection is modeled
on features in the world of the breeder that obviously help in reproduction
against fellow species members. Darwin spoke of male combat, as when
two stags fight for the harem and as a result the antlers are under strong
selection pressure to increase in size – modeling this on dog and cock
breeders selecting for fiercer fighters – and female choice, as when the
peahen chooses the male with the biggest display of tail feathers – modeling this on breeders selecting for more beautiful birds and other like
organisms. Later in this book, we shall take up sexual selection in some
detail. For now it is enough to note that it was always an integral part of
Darwin's thinking.

There were other things that Darwin talked about, including something he called the "principle of divergence," where he argued that organisms split into different groups because of the pressure to exploit different
ecological niches – in-between kinds are literally neither fish nor fowl, and
cannot do as well as specialists. This splitting led Darwin to his metaphor
of the history of life as being like a massive tree, with the past in the roots
and the present at the tops of the boughs.

The affinities of all the beings of the same class have sometimes been
represented by a great tree. I believe this simile largely speaks the truth.

The green and budding twigs may represent existing species; and those produced during each former year may represent the long succession of extinct species. At each period of growth all the growing twigs have tried to branch out on all sides, and to overtop and kill the surrounding twigs and branches, in the same manner as species and groups of species have tried to overmaster other species in the great battle for life … As buds give rise by growth to fresh buds, and these, if vigorous, branch out and overtop on all sides many a feebler branch, so by generation I believe it has been with the great Tree of Life, which fills with its dead and broken branches the crust of the earth, and covers the surface with its ever branching and beautiful ramifications. (129–30)

Darwin next moved over some problems he was not able to solve very satisfactorily, including the nature of heredity, what we today would call "genetics." We can skip this now because it will come up again shortly. He was now ready, for the rest of the *Origin*, to turn to the third part of his argument – "& chiefly from this view connecting under an intelligible point of view a host of facts." Basically what Darwin now did was to go through the whole range of the life sciences, looking at the problems facing researchers, and then offering solutions based on his central hypothesis of evolution through natural selection. Conversely, in a kind of feedback argument, he was using these solutions as evidence for the truth of his hypothesis. Just as a detective facing a murder will hypothesize that some unlikely person is the culprit and then turn to the clues – the bloodstains, the broken alibi, the motive, the method of attack – to convict, so Darwin turned to the clues of biology to establish the truth of his bold conjecture. I should say that he was not flying blind at this point, as it were, but that he was following the methodological prescription of the British historian and philosopher of science William Whewell (1840), who argued that such a type of explanation, what he called a "consilience of inductions," was just what is needed when you are trying to explain using a cause that no one sees and that may indeed be unobservable. Whewell was thinking in the context of the wave theory of light, where even though no one sees the actual waves it is accepted on circumstantial evidence such as the interference patterns in Young's Double Slit experiment. Darwin never thought we would ever see natural selection in action, but undeterred he set out to convince on the bloodstains, the broken alibis, the motives, of biology.

One of the first areas to which Darwin turned was that of instinct and social behavior. Like many of his contemporaries as well as those before him, Darwin was fascinated by the world of insects, particularly those that live in nests or hives, and he showed the power of his theory through selection of a beautiful example of what Richard Dawkins (1982) has called the "extended phenotype." Why is it that honeybees build hexagonal spaces for their young? Why not squares or circles or whatever? Through a number of rather ingenious experiments (involving the use of colored wax to see exactly how and when the bees use their building materials), Darwin was able to show that this is the most efficient use of the wax and makes for a structure as strong as you are ever going to get. He also spent some time comparing different groups of living insects, showing that there is a line of bees from those that make the crudest honeycombs to those that make the most complex and perfect, and from this he argued that we can as it were in one place and time see the chain through time that would have produced the hexagonal spaces that distinguish the most sophisticated insects today. Also, given that it is the social insects that yield some of the most wonderful examples of instinct, it is perhaps not surprising – although certainly a mark of his genius – that Darwin found himself wrestling with what is today known as the "levels of selection problem" (Brandon and Burian 1984). As is well known, honeybee workers (always female) are generally sterile and lay no eggs of their own. How can a process like selection, produced by the struggle for existence and reproduction, bring on something like this? Who benefits from the adaptations produced by natural selection? Is it always the individual or can it sometimes be the group? Later we shall look in more detail at this question, one that has been of much interest in recent years, including in the context of humankind. Suffice it to say here that Darwin had things of importance to say, although without a full theory of heredity he could not hope for a fully satisfactory answer.

Moving on to the fossil record, much of the time Darwin was on the defensive, trying to show why it is that there are so many gaps in the record. But then he started to make the positive points, particularly about the extent to which one finds earlier fossils in the record, fossils that look like the combination of very different extant organisms. Lying behind a discussion such as this was the kind of Germanic thinking that led the anatomist Richard Owen (1848, 1849) to his archetypal theory, where organisms within a group (like vertebrates) are seen as modifications of a basic ground plan or archetype, what the paleontologist (and popular science writer) Stephen Jay Gould was to call a *Bauplan* (Gould and Lewontin 1979). For an idealist like Owen, there was considerable doubt as to whether the archetype was a real organism, but for Darwin it was always an ancestor. It was this fact that led him to argue that his theory could take the Germanic thinking under its wing and show it to be derivative and not basic.

> It is generally acknowledged that all organic beings have been formed on two great laws – Unity of Type, and the Conditions of Existence. By unity of type is meant that fundamental agreement in structure, which we see in organic beings of the same class, and which is quite independent of their habits of life. On my theory, unity of type is explained by unity of descent. The expression of conditions of existence, so often insisted on by the illustrious Cuvier, is fully embraced by the principle of natural selection. For natural selection acts by either now adapting the varying parts of each being to its organic and inorganic conditions of life; or by having adapted them during long-past periods of time: the adaptations being aided in some cases by use and disuse, being slightly affected by the direct action of the external conditions of life, and being in all cases subjected to the several laws of growth. Hence, in fact, the law of the Conditions of Existence is the higher law; as it includes, through the inheritance of former adaptations, that of Unity of Type. (C. Darwin 1859, 206)

Another point which we can raise but not answer here is that of progress. Is the fossil record progressive, going from the simple to the complex, from the monad to the man? And what was Darwin's take on the issue? Again let us say that there are interesting and important questions here and that Darwin had interesting and important things to say on the matter.

Probably it was the variation among the denizens of the Galapagos Archipelago in the Pacific that set Darwin on the road to evolutionism. Hence, biogeography was always going to be a winner for him. Of course, there had to be a lot of discussion about how organisms could cross large oceans – Darwin always favored rafts and like phenomena rather than now-vanished land bridges – but basically it was a discussion that almost wrote itself, as Darwin showed how the oddities of geographical distribution fall away under the gaze of natural selection. The Galapagos naturally had a starring role, as did the fact that the organisms of islands tend to resemble their neighbors on local land masses rather than animals and plants on lands far away.

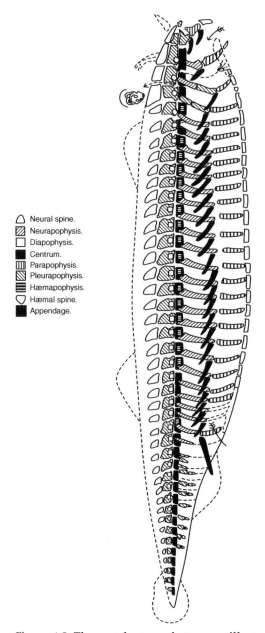

Figure 1.2 The vertebrate archetype, as illustrated by Richard Owen in
On the Nature of Limbs (1849).

The most striking and important fact for us in regard to the inhabitants
of islands, is their affinity to those of the nearest mainland, without
being actually the same species. Numerous instances could be given of
this fact. I will give only one, that of the Galapagos Archipelago, situated

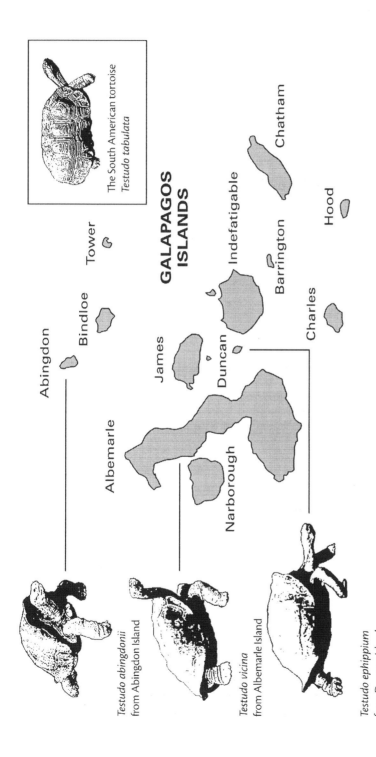

The South American tortoise
Testudo tabulata

GALAPAGOS
ISLANDS

Abingdon

Bindloe

Tower

James

Albemarle

Duncan

Narborough

Indefatigable

Barrington

Charles

Hood

Chatham

Testudo abingdonii
from Abingdon Island

Testudo vicina
from Albemarle Island

Testudo ephippium
from Duncan Island

Figure 1.3 Variation among Galapagos tortoises, showing how evolution occurred as they moved from the mainland and then from island to island. (These islands have been given their old names, as Darwin would have known them.)

under the equator, between 500 and 600 miles from the shores of South
America. Here almost every product of the land and water bears the
unmistakeable stamp of the American continent … Why should this be
so? why should the species which are supposed to have been created in
the Galapagos Archipelago, and nowhere else, bear so plain a stamp of
affinity to those created in America? There is nothing in the conditions
of life, in the geological nature of the islands, in their height or climate,
or in the proportions in which the several classes are associated together,
which resembles closely the conditions of the South American coast:
in fact there is a considerable dissimilarity in all these respects. On the
other hand, there is a considerable degree of resemblance in the volcanic
nature of the soil, in climate, height, and size of the islands, between
the Galapagos and Cape de Verde Archipelagos: but what an entire and
absolute difference in their inhabitants! The inhabitants of the Cape de
Verde Islands are related to those of Africa, like those of the Galapagos to
America. (C. Darwin 1859, 397–99)

Why should any of this be so? Obviously because of what Darwin called descent with modification. The denizens of the Galapagos came from South
America, the denizens of Cape de Verde came from Africa, and today they
show these different origins.

Darwin moved on through the fields. Systematics makes sense because
of common descent. In the eighteenth century, the Swedish taxonomist
Carl Linnaeus had put biological classification on a solid basis, a system
that we still use today. His genius was to invoke an ever-more comprehensive collection of nested sets (known as "taxa"), with seven levels (known
as "categories"). Every organism belongs to one (and only one) taxon at
each category level, and then at each higher level organisms are grouped
into ever larger taxa. This all makes good sense in the Darwinian system.

Naturalists try to arrange the species, genera, and families in each class,
on what is called the Natural System. But what is meant by this system?
Some authors look at it merely as a scheme for arranging together those
living objects which are most alike, and for separating those which
are most unlike; or as an artificial means for enunciating, as briefly
as possible, general propositions … But many naturalists think that
something more is meant by the Natural System; they believe that it
reveals the plan of the Creator; but unless it be specified whether order in
time or space, or what else is meant by the plan of the Creator, it seems

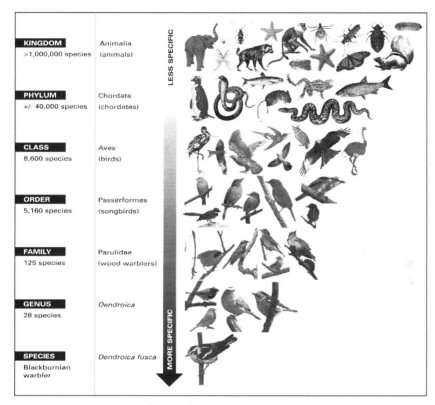

Figure 1.4 The Linnaean hierarchy.

> to me that nothing is thus added to our knowledge. Such expressions as
> that famous one of Linnæus, and which we often meet with in a more
> or less concealed form, that the characters do not make the genus, but
> that the genus gives the characters, seem to imply that something more
> is included in our classification, than mere resemblance. I believe that
> something more is included; and that propinquity of descent, – the only
> known cause of the similarity of organic beings, – is the bond, hidden as it
> is by various degrees of modification, which is partially revealed to us by
> our classifications. (413–14)

Anatomy likewise comes to life because of evolution through selection.
Phenomena noted and mysterious since the time of Aristotle – the homologies (as Richard Owen labeled them) between animals of very different lifestyle – are truly inexplicable if one is thinking purely in terms
of function. "What can be more curious than that the hand of a man,
formed for grasping, that of a mole for digging, the leg of the horse, the

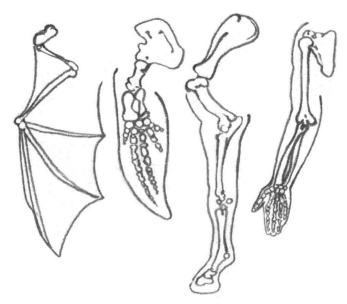

Figure 1.5 Homology between the forelimbs of mammals. What makes this example striking is that the similarities seem to have nothing to do with the different functions: the bat for flying, the porpoise for swimming, the horse for running, and the human for grasping.

paddle of the porpoise, and the wing of the bat, should all be constructed on the same pattern, and should include the same bones, in the same relative positions?" (C. Darwin 1859, 434). But of course they make perfectly good sense within the Darwinian picture. "The explanation is manifest on the theory of the natural selection of successive slight modifications, – each modification being profitable in some way to the modified form, but often affecting by correlation of growth other parts of the organisation. In changes of this nature, there will be little or no tendency to modify the original pattern, or to transpose parts" (435).

As this part of our discussion draws to a close, we can note that embryology was a particular point of triumph for Darwin. Everyone knew that the embryos of organisms very different – human and chick – frequently look very similar. How can this be? Descent with modification, obviously. But then Darwin went into detail, showing how selection and his analogy from the world of breeding can provide real insight. The Darwinian argument is that adults have been torn apart by selection in the struggle for existence. But the young, by and large, are protected from the struggle, so there is no reason to expect them to be made different by selection. Turning to the domestic world, Darwin hypothesized that since breeders are generally interested only in the adults, we should find that the young differ far less than their grown parents. Checking on horses – cart-horses versus racehorses – and dogs – greyhounds versus bulldogs – Darwin found his prediction to be correct, even though breeders had assured him that there was nothing to his supposition.

> Some authors who have written on Dogs, maintain that the greyhound and bulldog, though appearing so different, are really varieties most closely allied, and have probably descended from the same wild stock; hence I was curious to see how far their puppies differed from each other: I was told by breeders that they differed just as much as their parents, and this, judging by the eye, seemed almost to be the case; but on actually measuring the old dogs and their six-days old puppies, I found that the puppies had not nearly acquired their full amount of proportional difference. So, again, I was told that the foals of cart and race-horses differed as much as the full-grown animals; and this surprised me greatly, as I think it probable that the difference between these two breeds has been wholly caused by selection under domestication; but having had careful measurements made of the dam and of a three-days old colt of a race and heavy cart-horse, I find that the colts have by no means acquired their full amount of proportional difference. (444–45)

And so we come to the end of the *Origin*, with that final flowery passage about entangled banks, singing birds, flitting insects, and crawling worms. "There is grandeur in this view of life, with its several powers, having been originally breathed into a few forms or into one; and that, whilst this planet has gone cycling on according to the fixed law of gravity, from so simple a beginning endless forms most beautiful and most wonderful have been, and are being, evolved" (490).

After Darwin

What happened after the *Origin* was published? Did people accept Darwin's arguments? Did he, in the language of the philosophers, give us a new "paradigm"? The answer is somewhat complex (Hull 1973; Ruse 1979, 2009a). Nigh everyone knows that there was a major row, with Darwin's great supporter Thomas Henry Huxley debating Samuel Wilberforce,

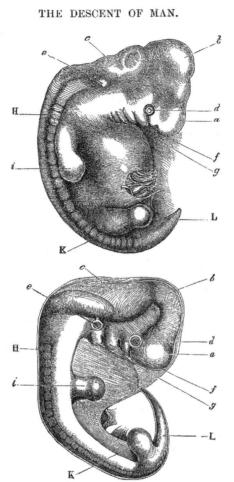

THE DESCENT OF MAN.

Fig. 1. Upper figure human embryo, from Ecker. Lower figure that of a dog, from Bischoff.

---

a. Fore-brain, cerebral hemispheres, &c.
b. Mid-brain, corpora quadrigemina.
c. Hind-brain, cerebellum, medulla oblongata.
d. Eye.
e. Ear.

f. First visceral arch.
g. Second visceral arch.
H. Vertebral columns and muscles in process of development.
i. Anterior }extremities.
K. Posterior }
L. Tail or os coccyx.

Figure 1.6 Embryological similarity between human and dog (from the *Descent of Man*).

bishop of Oxford. Supposedly Wilberforce (the son of William Wilberforce of anti-slavery fame) asked Huxley whether he was related to monkeys on his grandfather's side or his grandmother's side, to which Huxley replied that he would rather be descended from a monkey than from a bishop of the Church of England. Probably, as good stories so often are, this is apocryphal, but that there was opposition to evolution is beyond doubt. However, as far as the fact of evolution is concerned, very quickly – with the obvious exception of the American South (which in any case had other things on its mind as it entered on a civil war) – people accepted that it is true. That all organisms, living and dead, came about through a long, slow process of developmental change, or evolution as it was now being called. Even religious people came around by the end of the first decade after the *Origin* (say about 1870), agreeing that our origins, too, most probably lay in evolution. It is true that people (as today) generally wanted a special miracle for the insertion of immortal souls, but it was agreed that this is not a scientific claim anyway.

Natural selection, however, fared less well. No one wanted to deny it outright, but few thought it had the power that Darwin wanted to give it. There were a number of reasons for this rejection, and some were very solidly scientific. First, and most importantly, Darwin had no adequate theory of heredity (Vorzimmer 1970). But without such a theory, there is no guarantee that selection can ever be effective. Suppose a feature proves useful in the struggle for existence and the winner with the feature reproduces a lot. Unless one can be sure that the feature will be passed on, without too much dilution, success in one generation means nothing about the next generation. There cannot be the cumulative effect that the Darwinian process supposes. Second, there was the age-of-the-Earth problem (Burchfield 1975). The physicists, ignorant of the warming effects of radioactive decay, supposed that the Earth must be much younger than it truly is. Indeed, offering a figure of about 100 million years, with enough cooling to bear life coming only 25 million years ago, there seemed not to be time enough for a slow process like selection. No one really had much of an answer to this, even though now we know it was based on a mistake. Some people tried other mechanisms, like saltationism (evolution by large changes or jumps, as fox to dog in one generation). Darwin himself supplemented selection with more and more Lamarckism, the inheritance of acquired characteristics. Basically though, everyone knew that there was a big unsolved problem.

Almost paradoxically, given the 2,500-year history of opposition to evolution because of the problem of final causes, this supposed inadequacy of natural selection did not really worry a lot of the prominent biologists of the day, like Huxley. They were morphologists and paleontologists. They really only dealt with dead organisms, in the laboratory, as they cut them up and studied and classified them. Adaptation, so vital to a naturalist like Darwin, basically left them cold. They did not think it a very significant or interesting issue. Rather the evolutionists after Darwin tended to spend their days trying to ferret out phylogenies, the paths that organisms had taken as they evolved up to the present (Bowler 1996). This activity was reinforced and given meaning thanks to the fabulous finds now starting to pour forth from the American and Canadian Wests, as fossil hunters unearthed one incredible beast after another: *Allosaurus, Ceratosaurus, Brontosaurus, Camarasaurus, Amphicoelias, Diplodocus, Camptosorus,* and many, many others. Exciting stuff but basically not very conceptual, and although everyone used and praised Darwin's name, in major respects it was not really the science of the *Origin* – at least not with regard to causes.

All changed pretty rapidly around 1900. For a start, the physicists started to get on top of the idea of radioactive decay and it was realized that this has a warming effect on the Earth, and that it is in fact far older than thought previously. In fact, it is the decay itself that allows physicists to construct "clocks" that enable one to put absolute dates on events in the past. To use today's figures, the universe as a whole is around 15 billion years old, and the Earth and our solar system generally are about 4.5 billion years old. There is quite enough time for a slow process like natural selection, although as we shall soon see, it may well be that much selection-driven change is much quicker than Darwin believed.

At a more biological level, as is well known, it was also around then that researchers discovered and started to duplicate work done in the 1860s by a now-dead Moravian monk, Gregor Mendel. It was he who started to work out the true principles of heredity, making for a theory of genetics. Essentially he argued that physical features in any individual organism are controlled by factors of heredity (what we now call genes), that these come in pairs – one comes from the father and the other from the mother – and that these are passed on unchanged from generation to generation. Although it took some time for people to grasp the fact, this means that we now have a solution to Darwin's dilemma, because if an organism with a particular gene survives and reproduces, it will be this unchanged gene that goes on to the next generation, while the genes of the losers go nowhere. Selection can be effective (Provine 1971). People often regret that Darwin never knew of Mendel's work, thinking that this could have moved evolutionary studies along much quicker than actually happened. Perhaps, although more likely it would have made no difference. Mendel was working on technical questions in German biology and never thought of himself as providing an answer to Darwin's worries. In fact, he read a German translation of the *Origin* early in the 1860s and focused exclusively on the question of whether he, as a Catholic priest, could properly accept evolution. He decided he could!

By the second decade of the twentieth century, researchers led by Thomas Hunt Morgan in New York City had located the gene as something lying on the string-like entities (chromosomes) in the centers of cells, and also, vital for an evolutionary theory, they had realized that sometimes genes do change (mutate), yielding the physical or behavioral variations needed for the ongoing process of natural selection (Allen 1978). Then in the 1930s it all came together. First, mathematical biologists – population geneticists – showed conceptually how Darwinian selection and Mendelian genetics can work together. Following the labors of these men – Ronald A. Fisher (1930) and J. B. S. Haldane (1932) in Britain and Sewall Wright (1931, 1932) in the USA – the more empirical biologists went out and studied in nature and experimented in the laboratory, showing how one can build a fully causal and effective theory of evolution. Important were people like E. B. Ford (1964) and Julian Huxley (1942), the eldest grandson of Thomas Henry Huxley, in Britain and geneticist Theodosius Dobzhansky (1937), ornithologist and systematist Ernst Mayr (1942), paleontologist George Gaylord Simpson (1944), and botanist G. Ledyard Stebbins (1950) in America. All that was needed was the coming of molecular biology in 1953 and, after some initial hesitations, this was rapidly and completely integrated into the evolutionary mix and it was then possible to fulfill the vision of the *Origin*.

To complete this chapter, let me run now through modern evolutionary thinking trying to do two things. First, to show how very much in the Darwinian tradition modern evolutionary thinking really is. Second, to show how things have moved on. Nothing that is actually done by today's biologists is exactly (often, not even remotely) like the work of Darwin.

This is not a matter of despair or criticism. The best science never stands still, and this is certainly the case in the world of evolution.

Selection experiments

Darwin turned to the world of the animal and plant breeder for analogical evidence supporting natural selection. Evolutionists do the same today. For instance, one famous study that has been running for over a hundred years (it was started in 1896) is a selection experiment in Illinois on oil content in corn (maize) (Dudley 1977). Starting at around 4 to 6 percent, we are now up to 16 percent in the line selected for oil content. In the reverse line, oil content is almost zero. A somewhat more naturalistic experiment – meaning an experiment trying to parallel what was supposed to have happened in nature – was performed by geneticist Francisco J. Ayala (a student of Dobzhansky's) on fruit flies. Normally these little insects are killed by alcohol, but around wineries in California they thrive, living off the very substance that is usually fatal. Has selection brought this about? It seems that it has, because selection for survival in conditions of increasing alcohol content rapidly brought about tolerance – need even – of alcohol. Thanks to the tools of molecular biology, Ayala and associates were even able to trace the biochemical pathways that were involved in this newly acquired ability (McDonald *et al.* 1977).

Thomas Henry Huxley used to complain that a major reason why he was not a selection enthusiast was that no one had yet shown that selection can lead to reproductive isolation, the mark of speciation. Today, this worry has been addressed. One experiment started with a population of corn, mixed between yellow and white varieties, with no reproductive barriers between members of the group (Pasterniani 1969). Selection was for those plants that favored their own color for reproductive partners, that is white–white and yellow–yellow. Within five generations the two-color varieties were much more unwilling or unable to breed with plants that were not of their color. White–yellow and yellow–white liaisons were much reduced. There was nothing particularly mysterious about any of this. White was linked to early flowering and yellow to later flowering. Hence the two varieties were simply not pollinating, that is crossing, together. A species barrier was being erected.

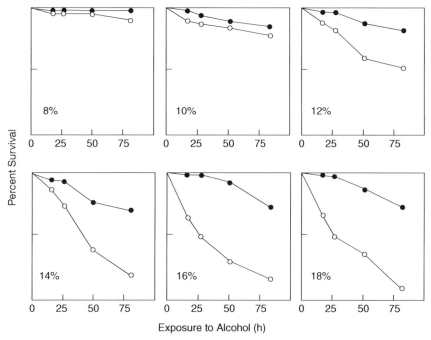

Figure 1.7 Survivorship of adult *D. melanogaster* flies from population selected for alcohol tolerance (•) and a population not selected (◦) in six different concentrations of ethanol. From McDonald *et al.* (1977).

The core arguments

Move on now to the second part of the Darwinian argument in the *Origin*. One thing that strikes the reader at once is that the modern theory is much more formal, much more mathematical, than the theory of Darwin. Traditional philosophy of science supposes that the best scientific theories are what are known as hypothetico-deductive (H-D) systems. These are axiom systems, rather like those you find in mathematics, with an initial group of unproven (within the system) premises and everything deduced from them (Hempel 1966). What distinguishes science from mathematics is supposedly that in the latter the premises (or axioms) are considered necessary in some sense, whereas in the former the premises are thought to be empirical. Newtonian mechanics, deducing the laws of Kepler and Galileo from the three laws of motion and the law of gravitational attraction, was considered the prime exemplification of such an H-D system. As it happens, in philosophical circles today it is generally less often thought that the H-D system is the aim of science. Rather one gets small systems or models, which are then applied to specific issues of interest (Beatty 1981). This is certainly the case with modern-day population genetics. This does not deny or belittle the fact that today's science tends to be heavily mathematical. It is just that scientists work a bit more like auto mechanics than the designers at head office. They are not trying to build grand systems, but to throw light on selected areas of empirical interest, and they just grab the tools to do this.

Formal or not, there is a major difference between Darwin's approach and post-Mendel approaches. Rather than trying to prove selection as basic, the ultimate law is the extension of Mendelism to groups, the Hardy–Weinberg law. This is an equilibrium law – basically it says that, all other things being equal – gene distributions will remain constant – and functions much as Newton's first law of motion functions, namely as a background, null condition against which change can be introduced and studied. Then the disruptive forces are introduced. These include migration in and out of a population, mutation, and selection. This does not mean that selection is any less important, but that it has a different conceptual position in modern evolutionary biology (Ruse 2008).

But is it no less important? It is no less important in the work of Fisher and Haldane, but it has a very different role in the thinking of Sewall Wright. His so-called "shifting balance theory" (SBT) owed less to the inspiration of Darwin and more to another mid-nineteenth-century British evolutionist, Herbert Spencer (Ruse 1996, 2004). Adaptation as such was important, but it was never the driving force for Spencer, and at a certain level this held true for Wright also. Most importantly, at the heart of the SBT there is the totally non-Darwinian force of genetic drift. Perhaps inspired by work on shorthorn cattle, Wright saw the greatest potential for change coming if groups are split into small sub-units (Provine 1986). In such units, it is quite possible for purely random encounters in mating to override the effects of selection. The non-adaptive genes can "drift" to dominate the group. But although these genes do not provide features that are themselves immediately adaptive (they have after all wiped out the genes that do provide such features), in other conditions it might be that these features of these randomly chosen genes are very advantageous. Perhaps therefore new features with

Figure 1.8 A "genetic landscape" showing the peaks (points of great fitness) and valleys (areas of least fitness). From Wright (1932).

the potential to be adaptive appear in some of these groups (and are able to appear only in such sub-units). Then, the small groups recombine into the large population again and, with conditions changed, the new, innovative features can now realize their potential and spread rapidly through the whole group. Note therefore that although selection takes over when the recombination occurs, in the sub-units it is all a function of random change, or drift. It is here that one gets the real change in evolution.

I hardly need say – and, if I do need say, Fisher said it repeatedly before me – that this is about as non-Darwinian a theory as it is possible to have. Breaking into small groups, drift, new variations that are adaptive appearing, and so forth – none of this is in the *Origin*. It simply is not. It is rather Spencerian (1862). He saw disruptive conflicting forces breaking equilibrium and then a scramble to regain order, usually at a higher level of organization in some sense. In Spencer's words – words picked up by Wright – "homogeneity leads to heterogeneity." In more detail, we have an initial state of balance, "dynamic equilibrium," disruption, regaining of equilibrium, with the conversion of homogeneity into heterogeneity (less complex into more complex). The genius of the younger American evolutionist was to translate all of this into the language of Mendelian genetics. Aided, I hasten to say, by one of the brilliant metaphors of evolution, the "adaptive landscape," which shows visually how genes climb up to the tops of adaptive "peaks," and yet how also they might find themselves in maladaptive "valleys."

If this is today's evolutionary biology, then I will be the first to proclaim the death of Darwinism. But is it today's theory? At the molecular level, it is now generally agreed that drift is very important. Most genes or stretches of DNA are adaptively neutral and their numbers drift up and down between being the total (100 percent) and being extinct (0 percent). This is the "neutral theory" of the Japanese biologist Motoo Kimura (1983). This, however, is not the physical level of organisms, about which Darwin was writing. Here it seems fair to say that the shifting balance theory has fared far less well. Even by the early 1940s, people like Dobzhansky (who had bought deeply into Wright's thinking) were starting to think that much genetic change, hitherto ascribed to drift, is in fact strongly under the influence of selection. Chromosome patterns in fruit flies, whose fluctuations were clearly linked to seasonal, climatic changes, were key evidence here. Drift was dropping in importance (Dobzhansky 1943; Ruse 1999). Today, no one seems to want to claim that drift is impossible or never occurs, but the shifting balance theory taken as a whole is generally dumped on the garbage pile of exhausted or false theories, like phlogiston before it (Coyne *et al.* 1997).

What about direct evidence for natural selection? I have suggested that Darwin himself was dubious about the possibility of finding this, believing that a selection in nature is always slow. He himself in the *Origin* gave fictitious examples, for instance of the evolution of fast and slow forms of predating wolves. A fellow Victorian naturalist, Henry Walter Bates (1862), one of the very few to take natural selection seriously, came up with a brilliant, selection-based explanation of butterfly mimicry. He showed in some detail, and with real experimental evidence, that because this gives them an adaptive edge, selection has caused some non-poisonous butterflies to mimic poisonous butterflies. Yet although Darwin was very excited – writing anonymous praising views of the work and even, through his publisher John Murray, getting Bates a job (as secretary of the Royal Geographical Society) – he introduced Bates's work in later editions of the *Origin* only toward the end (C. Darwin 1872, 375). Darwin thought it important, but not that important.

It seems in fact that Darwin missed an even greater opportunity of recording selection-driven change observable in our lifetimes. Today, the classic case of such change is so-called "industrial melanism" (Majerus 1998). It has been recorded over and over again that butterflies and moths, heavily predated by birds, have changed color as a function of the darkening of the trunks of trees (and other vegetation) thanks to the pollution from industry. A correspondent, Albert Brydges Farn, in a letter dated November 18, 1878, pointed this fact out to Darwin himself!

The Dartons, Dartford, Kent.
18th. November, 1878.

The belief that I am about to relate something which may be of interest to you, must be my excuse for troubling you with a letter.

Perhaps among the whole of the British Lepidoptera, no species varies more, according to the locality in which it is found, than does that Geometer, Gnophos obscurata. They are almost black on the New Forest peat; grey on limestone; almost white on the chalk near Lewes; and brown on clay, and on the red soil of Herefordshire.

Do these variations point to the "survival of the fittest"? I think so.

It was, therefore, with some surprise that I took specimens as dark as any of those in the New Forest on a chalk slope; and I have pondered for a solution. Can this be it?

It is a curious fact, in connexion with these dark specimens, that for the last quarter of a century the chalk slope, on which they occur, has been swept by volumes of black smoke from some lime-kilns situated at the bottom: the herbage, although growing luxuriantly, is blackened by it.

I am told, too, that the very light specimens are now much less common at Lewes than formerly, and that, for some few years, lime-kilns have been in use there.

These are the facts I desire to bring to your notice.

I am, Dear Sir,
Yours very faithfully,
A. B. Farn

C. R. Darwin Esq. F.R.S. &c &c &c (Darwin Correspondence Project Database, www.darwinproject.ac.uk/entry-11747 [letter no. 11747])

It seems that Darwin never replied and that he did not pick up on the significance of Farn's information.

Today, industrial melanism apart, the most famous example of a study of evolution in action is that of Rosemary and Peter Grant, on the finches of the Galapagos (P. R. Grant 1986; B. R. Grant and P. R. Grant 1989; P. R. Grant and B. R. Grant 1995; 2007). For over thirty years now, they have been studying – counting and measuring – the birds on a small islet, Daphne Major, looking particularly at beak size and shape. They have shown beyond reasonable doubt that in times of drought the beaks get bigger and stronger, the better to crack and chew up nuts and cactus, and in times of rain the beaks get smaller and finer, the better to handle small seeds and the like. The Grants have counted the living and the dead, and they truly show that these changes are not chance but a function of differential reproduction, that is of natural selection.

The Grants are not alone in showing how Darwin quite underestimated the rate of evolution in the wild. There are many studies on animals, on plants, on micro-organisms demonstrating just this (Endler 1986). Some of the most interesting and persuasive – which we shall touch on in the final chapter – are of parasites and pests that evolve rapidly in response to human efforts to eliminate disease and destruction. In short, therefore, both theoretically and empirically today's evolutionary biology is very much in the spirit of Darwin's work (as opposed to his contemporaries, who doubted his causal achievements) but equally goes very much beyond it – as one would expect in a fruitful and forward-looking science.

It comes as no surprise that the tree of life today is very much more filled out than it was in Darwin's day (Ayala 2009; Woese 1998). For a start, thanks particularly to molecular techniques, we know that the branch of greatest interest to us – that which includes the animals and plants – is only one of three major divisions. We belong to the Eukarya (organisms with complex cells), as distinguished from the Bacteria and the Archaea (organisms with more simple cells). We also know that it is possible for genetic material to be transferred laterally across lines, thanks to the actions of viruses, which can take pieces of material (a nucleic acid molecule) and having incorporated these into themselves, then infect another organism and pass the material on. It seems clear that this is an important process with the simpler organisms, but how significant it is as life gets more complex seems still to be a matter of debate. Of course, if "horizontal gene transfer" gets ever-more important, although increasingly this would wreck a simple picture of a tree of life, it would not challenge the importance of natural selection.

The consilience today

Third and finally let us turn to the last part – actually about the last three-fifths – of the material covered in the *Origin of Species*: the consilience. Simply put, this form of argumentation is as crucial today as it ever was. This is not to say that there is no change. There is massive change. It is to say that the structure of the *Origin*, including here the intention to explain through natural selection, is (notwithstanding the cries of critics) embraced fully by today's evolutionary biologists. Let us go quickly through the sub-branches or disciplines.

First there is instinct, by which Darwin very much meant social behavior. In the past half century, this whole topic has exploded outwards, in theory and in experiment and in study in nature. It has even been given its own name of "sociobiology." The key conceptual moves came in the 1960s and early 1970s, thanks to British evolutionary biologists, notably William D. Hamilton (1964) and John Maynard Smith (1982), and American evolutionary biologists George C. Williams (1966) and Robert Trivers (2002). Rapidly, building on the conceptual insights, there was massive empirical attention and, by the middle of the 1970s, the world's leading ant biologist, Edward O. Wilson of Harvard, was able to write his magnificent overview *Sociobiology: The New Synthesis* (1975) and Richard Dawkins of Oxford penned his popular account *The Selfish Gene* (1976).

Take a first-class exemplar of this kind of work, Wilson's own study of the leaf-cutting ants of the Amazon, the *Atta*. They have many castes – the large soldiers, the smaller foragers, and then the leaf-cutters, the gardeners who tend the growing of fungi on the chewed-up leaves, the nursery workers who tend the young, and of course the massive queen. All told, the ant genus *Atta* has seven classes of workers.

> A key feature of *Atta* social life … is the close association of both polymorphism and polyethism with the utilization of fresh vegetation in fungus gardening … An additional but closely related major feature is the "assembly-line" processing of the vegetation, in which the medias cut the

vegetation and then one group of ever smaller workers after another takes the material through a complete processing until, in the form of 2-mm-wide fragments of thoroughly chewed particles, it is inserted into the garden and sown with hyphae. (E. Wilson 1980a, 150)

Why the different castes and why the proportions? Wilson ran experiments (usually involving removing whole castes) to determine the most efficient way of using resources. If the ants want to get the most bang for their buck, less metaphorically get the greatest results in the sense of producing new, fertile gene bearers for the minimum amount of effort, what role do the various castes play and is the proportion of one to another the most efficient? This kind of "optimality" thinking (Oster and Wilson 1978) – what would natural selection do to achieve the greatest adaptive efficiency? – paid big dividends. "What *A. sexdens* has done is to commit the size classes that are energetically the most efficient, by both the criterion of the cost of construction of new workers … and the criterion of the cost of maintenance of workers" (E. Wilson 1980a, 150). I hardly need say that this is way beyond any kind of thinking that Darwin had, and yet in another sense is completely Darwinian. Consider Darwin in the *Origin*, speaking of sterile workers: "we can see how useful their production may have been to a social community of insects, on the same principle that the division of labour is useful to civilised man" (C. Darwin 1859, 241). It is this division of labor that is central to Wilson's thinking. You have different castes in the proportions that you do because to have everyone doing everything would simply be biologically inefficient.

The fossil record always fascinates. Again, in some respects, we are way beyond Darwin. As noted, we have the ability now to assign absolute dates. It is believed that life itself started about 3¾ billion years ago, that is to say, just about as soon as it might have done given the need of the initially molten Earth to cool to livable temperatures. The Cambrian, that explosion of complex life (including things like the trilobites) that started things on the way to the vertebrates and so eventually to us, began about 540 million years ago. It is after this that the fossil record really takes off. To his embarrassment and the delight of his critics, Darwin had no fossil evidence of any earlier organisms. Life seems to have started entire and complex, in one fell swoop. Countering, he gave text-book examples of scientific adhockery – the fossils would be where the oceans are now and, even if we could drill, the weight above would have squashed them into non-being. Fortunately now none of this is necessary. We have a pretty good record of pre-Cambrian life from its beginnings and moreover this is life which starts simple and gets more complex, precisely as expected (Knoll 2003). We have lots more transitional fossils, most recently the fish-amphibian *Tiktaalik* from the upper reaches of snowy Canada.

Much contested in recent years is the question of whether Darwin was right in seeing (as he did) the history of life as a smooth process, leading from one form to another. For Darwin, this was part and parcel of his commitment to adaptation. Rapid changes would take organisms out of adaptive focus. He had no place for "hopeful monsters." Famously, the late Stephen Jay Gould together with fellow paleontologist Niles Eldredge argued that the true course of history is much more one of stop and go – periods of relative evolutionary inaction (stasis) broken by times of rapid change (Eldredge and Gould 1972). Much ink has been spilt over this theory of "punctuated equilibrium," including the question of how smooth a Darwinian gradual change must necessarily be. Ultimately, it is probably true that, as with so many controversies, there is some truth in the Gould–Eldredge hypothesis but nothing like as much as was claimed at times by enthusiasts (Sepkoski and Ruse 2009). Rates of change do vary, but then what would you expect? At the micro level, change is probably going to be pretty gradual, although precisely what one might mean by "pretty gradual" can be contested. Dimensions might be distorted drastically, very quickly, with major implications. Richard Dawkins's (1996) example is of a 747 jumbo jet being stretched, virtually overnight, from its original form to make room for more passengers. We might get a change in (say) reptilian form pretty quickly this way. It is, however, unlikely that we would get something like the overnight change from reptile to bird. And of course the bird-reptile *Archaeopteryx* shows that this did not happen. (The *Archaeopteryx* fossils started to emerge in the early 1860s, and soon found their way into the *Origin*.)

The consilience continued

Biogeography was one of Darwin's strongest supports for his theory and it continues to be so today. But, because we are now in the era of plate tectonics, leading to continental drift, here the changes have been truly staggering. A major question for Darwin and his contemporaries was how organisms might actually travel around the globe, and especially how they might journey across large open tracts of water. Darwin spent much effort looking into how decaying vegetation could surf the ocean waves and how seeds could attach themselves to the feet of migratory birds. Others supposed bridges and causeways that no longer exist. However, although no one denies the powers of driftwood and the like, there is no need of fabulous hypotheses – and especially not fabulous hypotheses about now-missing land links – because we now know that the continents move around the globe on massive plates, things which arise out of the Earth, move in a stately fashion across the surface, and finally return into the bowels in an almost Wagnerian fashion. You can explain for instance why it is that the Permian reptile *Lystrosaurus*, a slug of a brute if ever there was one, can be found in Africa, India, and Antarctica. It did not go traveling of its own volition. It stayed where it was and let the moving plates do the hard work (Ruse 2008).

Systematics, morphology, embryology – they start to bring the consilience to a close. One could write full-length essays on each of these topics. Change and yet adaptation as a result of natural selection is the story. Systematics was transformed in the 1970s, thanks to the coming of cladistics, the system of classification based on the somewhat idealistic system of the German biologist Willi Hennig (1966). Although there were some (rather extreme devotees of the philosophy of Karl Popper) who questioned whether classifications need at all reflect history, generally it was (and still is) agreed that cladistic classification is firmly evolutionary. It is about paths or phylogenies. Less obvious, particularly at first, was whether it was always very Darwinian. For instance, change in a line without branching would not be (could not be) recorded. As methods have become more sophisticated and refined, however, particularly thanks to the coming of molecular techniques, it does seem that the gap between cladistic practices and results and Darwinian processes has narrowed.

As it happens, like paleontology, morphology has been very controversial and for much the same or at least related reasons. Stephen Jay Gould argued that form, as in homology, is basic, and function, as in adaptation, comes afterwards. Certainly (according to him and co-writer Harvard geneticist Richard Lewontin) it does in many cases. Most particularly, because form puts its mark on organisms, much that we find has little or no relation to utility. Often, seemingly important adaptations are "spandrels,"

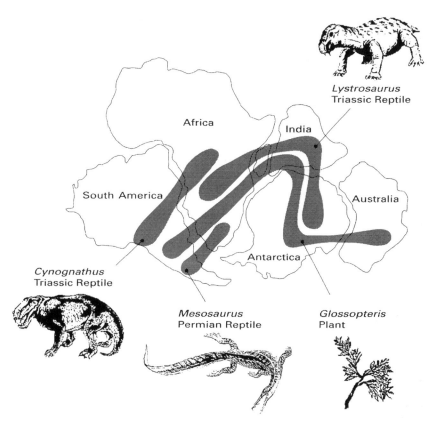

Figure 1.9 The effects of continental drift on the distributions of (the fossil remains of) animals and plants.

that is to say by-products of other characteristics, which may or may not themselves be adaptive. Spandrels are the triangular spaces at the tops of columns in medieval buildings, frequently, as in the church of San Marco in Venice – with beautiful decorations all over them. "The design is so elaborate, harmonious, and purposeful that we are tempted to view it as the starting point of any analysis, as the cause in some sense of the surrounding architecture." But this is to reverse cause and effect. "The system begins with an architectural constraint: the necessary four spandrels and their tapering triangular form. They provide a space in which the mosaicist worked; they set the quadripartite symmetry of the dome above" (Gould and Lewontin 1979, 148). Warming to his theme, arguing that much of the living world is non-adaptive, Gould sneered that those who seek adaptation are too often like Dr. Pangloss in Voltaire's *Candide*,

Figure 1.10 Spandrel from the Villa Farnesina, Rome, painted by Raphael and co-workers, 1511.

forever seeing value or use when there is none – "the best of all things in the best of all possible worlds." According to Gould, Darwinians spin "just so" stories, at one with the tales of Rudyard Kipling, who tells us that the elephant has a long trunk because a crocodile pulled it!

Take all of this with a pinch of salt. That the organic world is marked by adaptive complexity is, despite Gould's rhetoric, simply a common-place. Since Darwin, again and again the most outlandish feature has been shown to be tightly tied to reproductive utility. Take the triangular plates running along the back of the *Stegosaurus*, a monstrous dinosaur discov-ered in the American West in the decades after the *Origin*. What can they be for? Some have suggested sexual attraction; others have proposed that their utility comes as weapons fighting or defence. The popular view today is that they are for temperature control, for heating and cooling (Farlow *et al.* 1976). The brute needed to warm up its blood first thing in the morning and then, at midday, particularly with the heat produced by its herbivorous diet, it needed to get rid of the heat. The plates are just like the fins one finds in electrical cooling towers, where they are similarly used for heat transference. Supporting the hypothesis, the fossil evidence is that the blood was moved from the main body (and back again) in just the efficient way one would suppose were heat transference the goal in mind.

It is no less a commonplace for the Darwinian, starting with the *Origin* itself, that not every feature is going to be adaptive. Introducing the idea of natural selection, Darwin concluded: "Variations neither useful nor injurious would not be affected by natural selection, and would be left a fluctuating element, as perhaps we see in the species called polymorphic" (Darwin 1859, 81). There are many reasons why features might not be very adaptive or might even be positively non-adaptive. In certain respects the peacock's tail is very adaptive, that is when it comes to attracting the peahen, but is very non-adaptive when it comes to escaping from predators. Walking upright has many adaptive virtues for humans, not the least the way our hands are freed for other tasks. But the cost has been a less than well-designed lower back, as many sufferers attest. The appendix may have had a role once, but is of questionable or possibly alternative value now, and for some people it is a thing with a significantly negative value. These and other points are considered in more detail later.

As sociobiology has transformed the study of instinct, so the new field of evolutionary development has transformed embryology. The molecular biologists have moved in, giving new insights and empirical discoveries, and offering new theories. This is no bad thing, if only in the sense that the neo-Darwinian synthesis of the 1930s followed the population geneticists in regarding development as something of a black box – you have genes (genotypes) on the one side, and you have physical features (phenotypes) on the other, and no one cared much about what went on in-between. Pigs in, sausages out, and no questions please. Now this connection is brought into the sunlight, as biologists trace in detail the paths from the nucleic acids to the performing physical organisms. Some find the experience so giddy-making that they are all for throwing out the old completely.

> The homologies of process within morphogenetic fields provide some of
> the best evidence for evolution – just as skeletal and organ homologies
> did earlier. Thus, the evidence for evolution is better than ever. The
> role of natural selection in evolution, however, is seen to play less
> an important role. It is merely a filter for unsuccessful morphologies
> generated by development. Population genetics is destined to change if it
> is not to become as irrelevant to evolution as Newtonian mechanics is to
> contemporary physics. (Gilbert *et al.* 1996, 368)

Well, yes. Certainly the first claim is true. Ernst Mayr, writing in 1963
about homology, having described it as the best of all kinds of evidence for
evolution, warned non-biologists not to look for impossible isomorphisms
between organisms as different as humans and fruit flies. You are not
going to find them. One hopes that wherever he is, looking up or looking
down, Mayr now celebrates the fact that at the molecular level there are
the most incredible homologies between humans and fruit flies. It turns
out that the genes that control development, the Hox genes, are nigh identical between the two organisms – and with many others too (Carroll *et
al.* 2001). But what about the second claim? Why any of this should impact
negatively on natural selection is hard to imagine. We now know that
organisms are built on the Lego principle – the same building blocks can
make the White House or King Kong, a fruit fly or a human. But natural
selection has no less of a role. Going back to earlier discussion, it is true
that "evo-devo" shows that often you can get fantastic changes by altering
what you have rather than by starting anew – the stretch 747 – but selection is no less important – if the plane does not fly, then it is of no use.
From Thomas Henry Huxley on, there has been a downplaying of adaptation, especially by bench biologists who never see organisms alive and
well, in their native habitats. But organisms do flourish, in their native
habitats, and if biologists forget this fact, if they forget it is one thing to
make an organism but then it must succeed in life's struggles, they will
never get the full story of evolution.

The consilience lives! We began this chapter with the pre-Socratics,
let us end with them. Heraclitus said that you cannot step into the same
river twice. Everything changes. Parmenides denied the possibility of such
change. All motion is delusion. As the greater philosopher Plato showed,
both were right in their respective claims and wrong in thinking that
theirs alone was sufficient. The world does change, things like mathematics

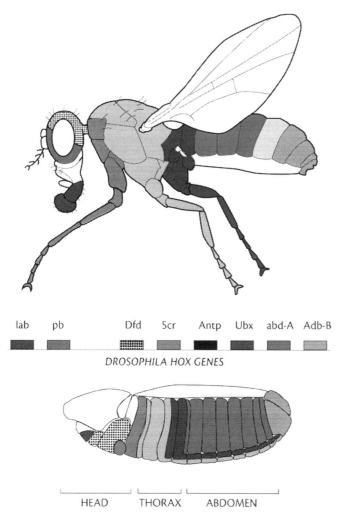

lab   pb   Dfd   Scr   Antp   Ubx   abd-A Adb-B

DROSOPHILA HOX GENES

HEAD   THORAX   ABDOMEN

Figure 1.11a The adult fruit fly is shown at the top, the embryo at the bottom, and the Hox genes in the middle. These genes sequentially order the manufacture of the bodily parts.

Fly Dfd	PKRQRTAYTRHQILELEKEFHYNRYLTRRRRIEIAHTLVLSERQUKIWFQNRRMKWKKDN	KLPNTKNVR
AmphiHox4	TKRSRTAYTRQQVLELEKEFHFNRYLTRRRRIEIAHSLGLTERQIKIWFQNRRMKWKKDN	RLPNTKTRS
Mouse HoxB4	PKRSRTAYTRQQVLELEKEFHYNRYLTRRRRVEIAHALCLSERQIKIWFQNRRMKWKKDN	KLPNTKIRS
Human HoxB4	PKRSRTAYTRQQVLELEKEFHYNRYLTRRRRVEIAHALCLSERQIKIWFQNRRMKWKKDN	KLPNTKIRS
Chick HoxB4	PKRSRTAYTRQQVLELEKEFHYNRYLTRRRRVEIAHSLCLSERQIKIWFQNRRMKWKKDN	KLPNTKIRS
Frog HoxB4	AKRSRTAYTRQQVLELEKEFHYNRYLTRRRRVEIAHTLRLSERQIKIWFQNRRMKWKKDN	KLPNTKIKS
Fugu HoxB4	PKRSRTAYTRQQVLELEKEFHYNRYLTRRRRVEIAHTLCLSERQIKIWFQNRRMKWKKDN	KLPNTKVRS
Zebrafish HoxB4	AKRSRTAYTRQQVLELEKEFHYNRYLTRRRRVEIAHTLRLSERQIKIWFQNRRMKWKKDN	KLPNTKIKS

Figure 1.11b Comparison of the proteins produced by the Hox genes of fruit flies and vertebrates. Note the very close homology between flies and humans. From Carroll *et al.* (2001).

are eternal. There is a truth here also when we come to Darwinian theory. Everything has changed. Not one item of Darwin's *Origin* has gone unmodified or unreplaced. Nothing has changed. We still have natural selection, analogies from human-run experiments, the consilience going successfully through all the branches of the life sciences, and much, much more. But we have said enough for now about the basic picture. We shall add to and qualify our understanding as we proceed. Now the time has come to turn to that ever-fascinating organism *Homo sapiens*.

2 Human evolution

And God said, Let us make man in our image, after our likeness: and let them have dominion over the fish of the sea, and over the fowl of the air, and over the cattle, and over all the earth, and over every creeping thing that creepeth upon the earth.

So God created man in his own image, in the image of God created he him; male and female created he them.

This is the story of human creation as given in Genesis (1:26–27), and with the coming of Christianity it was the story that was universally accepted for over 1,500 years. As noted in the first chapter, St. Augustine cautioned that one must be careful with demands that the Bible be interpreted literally. The story of human origins as given in Genesis shows the need of such a policy, for in the second chapter we get a rather different account, where woman is created as something of an afterthought to keep man happy and busy.

And Adam gave names to all cattle, and to the fowl of the air, and to every beast of the field; but for Adam there was not found an help meet for him.

And the LORD God caused a deep sleep to fall upon Adam, and he slept: and he took one of his ribs, and closed up the flesh instead thereof;

And the rib, which the LORD God had taken from man, made he a woman, and brought her unto the man.

And Adam said, This is now bone of my bones, and flesh of my flesh: she shall be called Woman, because she was taken out of Man. (Genesis 2:20–23)

The word from people like St. Augustine was that Christians should not spend their time trying desperately to reconcile texts like this but rather to spend time understanding the metaphorical or allegorical meanings that lie behind the words. The same is true when the text seems to conflict with up-to-date understanding in science and philosophy. The Bible was not written as a source of empirical information, but about God, humans, and their relationship. This does not mean that anyone doubted the Adam and Eve story – and there were of course good theological reasons for taking it fairly literally – but it does mean that the Christian tradition had within it the tools to work with science as and when the facts and theory demanded.

Famously the Irish archbishop James Ussher had (in 1650) announced, on the basis of the genealogies in the Bible, that the Earth is less than 6,000 years old, commencing on October 23, 4004 BC (starting on the sunset of the evening before, actually). But even then there were those who were, on the basis of ancient Egyptian and similar records, challenging such a short time span (Livingstone 2008). It was Isaac La Peyrère, a French Calvinist of Portuguese Jewish origins and author of *Prae-Adamitae* (Men before Adam), published in 1655 (and existing in manuscript for some years previously), who really put the cat among the pigeons, or rather the ancestors before Adam. Criticized so often that he clearly had a wide and enthusiastic readership, La Peyrère argued, on the basis of a somewhat convoluted reading of various sentiments expressed by St. Paul on the nature of sin and law, that there were clearly men before Adam and that these were not necessarily very nice. These people were not Adam's ancestors – Adam apparently was the father of only the Jewish race – so La Peyrère was what became known as a "polygenist" (meaning he thought there were multiple origins of the human race, as opposed to "monogenists" who think there is only one origin). La Peyrère made much of the fact that his position gave a ready answer to some of the biggest biblical conundrums, particularly those folk in the land of Nod, among whom Cain was supposed to have dwelt and found a wife. He also threw in some views about the local nature of Noah's flood and speculations that it was having relations with non-Jews that so upset the Lord that he found it necessary to wipe out almost every descendant of Adam.

Obviously none of this is very scientific. In fact, it isn't scientific at all! But it did start people thinking the unthinkable, namely that Genesis might not be the last (or the first) word on human origins. As we saw in the last chapter, it was in the eighteenth century that evolutionary ideas started to emerge, and as we also saw (if not discuss) that humans were a central item in these visions and speculations, namely the organisms right at the top. For evolutionists, human ancestors came with the territory. What we also saw was that early evolutionism was more a matter of ideology than of hard, empirically based reasoning. That was not to come until the nineteenth century. However, even in the eighteenth century some people did start to move toward the empirical. Noteworthy was Linnaeus, whom we have seen as the father of the modern system of organic classification. He put humans firmly into the animal picture, classifying us along with the sloth (!) and the ape. How far Linnaeus himself was inclined to evolutionary ideas is still a matter of debate. What is important is that in significant respects, for all that he stressed our spiritual and intellectual nature, he located us in the natural, animal world. Of a somewhat different order, but also important, was the fact that increasingly travelers were bringing back reports of human-like animals, what we today would probably put in the higher apes category. Fascinating is the description that the great French naturalist Georges Leclerc, comte de Buffon, gave of the chimpanzee, although he called it a Jocko or "smaller orang." Note how it is pictured as far more human-like than it really is. Again, like Linnaeus, Buffon was happy to stress the intellectual and spiritual side to humankind, separating us from the brutes, but the fact is that physically we are awfully close (Greene 1959).

At the beginning of the nineteenth century, the great French comparative anatomist Georges Cuvier (1813) – he who stressed the conditions of existence that Darwin thought so crucial to understanding organisms – was always opposed to the concept of evolution. Yet he shared with his arch-rival the evolutionist Lamarck the conviction that humans come late and at the top (Coleman 1964). Although he was a practicing Protestant (he was born in a border state that was as much German as French), Cuvier would never have based a scientific claim on biblical evidence, except insofar as the Bible could be considered a reliable historical record. He was the first really to see that the fossil record is roughly progressive and, finding no evidence of humans down in the record, concluded that we must be very recent arrivals on the scene. What was of great importance to Cuvier was the growing conviction that some organisms of the past no longer exist today. Could it have been that life existed and then vanished before humans arrived? This was more what is known as a natural theological

Figure 2.1 Buffon's "Jocko." From Georges Louis Leclerc, comte de Buffon (1707–88), *Histoire naturelle, générale et particulière*. Deux-Ponts: Sanson, 1785–91.

worry (that is to say, one based on reason and the senses) than a revealed theological worry (one based on faith and the Bible). But it was troublesome nevertheless. Working at this problem, general opinion tied things in with the common belief that the Earth was cooling and that God was getting the world ready for us, and that earlier life could only exist at the then-higher temperatures. God simply didn't want good living space to go to waste.

By this time, as must be clear, no one taking science seriously was confining speculations to a biblically based age of the Earth. Some determined on a reconciliation supposed that there were long gaps in the record unrecorded in the Bible. Others simply supposed that the six days were truly six very long periods of time. After all, the Bible did say that a thousand years are but as a day in the eye of the Lord. What became increasingly clear, as the century went on, was the fact that even though humans may be young, we are not that young. In the 1830s, the French geologist Boucher de Perthes had uncovered flint instruments of great antiquity. At first his discoveries were challenged and disregarded, but by the 1850s sentiment increasingly inclined to accept that he had made significant finds, showing that humans lived in France back when it was warm enough to support elephants and rhinoceroses. Indeed, there was evidence suggesting that humans may well have lived back when there were animals that have now gone extinct.

Enter Charles Darwin. It is notorious how brief was the discussion of humankind in the *Origin*. He left the topic until virtually the end and then dropped in what must be the most understated paragraph of the nineteenth century.

> In the distant future I see open fields for far more important researches.
> Psychology will be based on a new foundation, that of the necessary
> acquirement of each mental power and capacity by gradation. Light will
> be thrown on the origin of man and his history. (C. Darwin 1859, 488)

But no one was deceived. Everyone knew that this was the crucial piece in the story. What about us? What about *Homo sapiens*?

One should not be misled by Darwin's understated paragraph. He himself had absolutely no doubts that humans are completely and entirely a part of the natural world, the living world of plants and animals. Back in 1831, the *Beagle* left England carrying three native people from the bottom of the Earth, the tip of South America, Tierra del Fuego. They had been brought to England by the captain on a previous trip and, having been "civilized," were now being returned along with a missionary, in order to start the process of bringing their fellow "savages" (as everyone including Darwin would have termed them) into the modern world. This project ended in total failure, with the Anglicized natives rapidly returning to their original primitive state and the missionary having to be rescued before he could make a single convert. The ship's naturalist learnt a lesson that he never forgot. The most advanced of us is far closer to an animal state than anyone could possibly imagine. The veneer of civilization is thin indeed. Hence the reticence in the *Origin* was no function of indecision. Nor was it one of cowardice. Quite the contrary. It was very much part and parcel of the skilled presentation of the basic ideas that we saw in the previous chapter. Darwin knew full well that everyone would be obsessing about humans, but first he wanted to get the basic theory on the table, as it were. Hence the silence on our species. At the same time, he did not want to conceal his own thinking. Hence the brief paragraph at the end of the book.

Darwin was right in thinking that the human issue would be front and foremost. We have seen already that it was our species that Thomas Henry Huxley and the bishop of Oxford were jousting over, and this was but a part of a larger ongoing debate that Huxley carried on with the leading anatomist Richard Owen, over whether the human brain has unique parts (as Owen claimed) or whether there is nothing in our brain that cannot also be found in the brains of higher apes (as Huxley claimed) (Rupke 1994). This debate gave Huxley the excuse to write a whole book on humans and evolution – *Evidence as to Man's Place in Nature* (1863). He had to admit to a large jump from the highest apes to humans, but this did not stop him using a frontispiece that put humans in line with the apes, or from pointing out that the gap between us and the apes is less than the gap between the highest and the lowest apes. Of course Huxley had no solid fossil evidence to back up his arguments, although by this time people did know of the existence of a primitive kind of human from the Neander valley in Germany. (They are known as "Neanderthals," after the German word for valley – somewhat confusingly, at the beginning of the twentieth century, Germans changed the word from "Thal" to "Tal" and so more common today is "Neandertal".) These were discovered in 1856, three years before the *Origin* was published, although it does seem clear that specimens had been found elsewhere, earlier in the century. Opinion was divided over whether this was a new species or merely a subspecies of our species, *Homo sapiens*. Huxley inclined to the latter view. There were some indeed who thought that Neandertals are still with us, and suggestions were made that those interested might contemplate a trip to the west coast of Ireland.

If left to his own devices, I am not sure whether Darwin would ever have written a whole book on humankind. After the *Origin*, he seems to have been much happier fiddling around with little projects that caught his attention – the breeding devices of orchids, the ways of climbing plants, the effects of earthworms. But he was dragged into the debate by the co-discoverer of natural selection, Alfred Russel Wallace. In his early

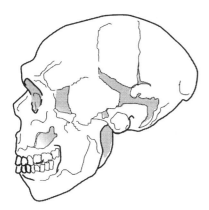 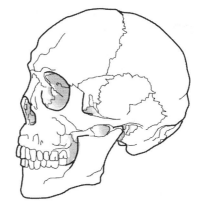

Figure 2.2 *Homo sapiens neandertalensis.*   *Homo sapiens sapiens*

life, Wallace seems to have had little or no religious belief. Sometime in the 1860s, by which time he had returned from the Far East and settled in England, Wallace became (as did many others) enthused by spiritualism (Wallace 1905). He started to believe in unseen forces and that these guide important events on Earth. In particular, from endorsing an entirely naturalistic view of human origins, he became convinced that only a kind of divine power could have produced *Homo sapiens* (Wallace 1870). In support, he instanced things like our great intelligence and our hairlessness, neither of which he thought could have been produced by natural selection. Reasonably, Wallace – who had indeed lived with natives – pointed out that such people really do not use all of their intelligence and so it is hard to see why it would have been cherished by natural selection.

Darwin was in despair. He thought, with some good reason, that Wallace was driving a stake into the heart of their theory. Yet at the same time he agreed that Wallace made good points about the difficulty of seeing how natural selection could have been responsible for many human features. It was at this point that he turned to sexual selection. Could what natural selection could not do perhaps be done by sexual selection? Hairlessness and so forth, intelligence even, were the end results of selection by males for females and (perhaps) conversely. I should warn, however, that having made this general move, in filling in the details Darwin did rather let his Victorian imagination run wild.

It is well known that with many Hottentots women the posterior part of the body projects in a wonderful manner; they are steatopygous; and Sir

Andrew Smith is certain that this peculiarity is greatly admired by the
men. He once saw a woman who was considered a beauty, and she was
so immensely developed behind, that when seated on level ground she
could not rise, and had to push herself along until she came to a slope.
Some of the women in various negro tribes are similarly characterised;
and, according to Burton, the Somal men "are said to choose their wives
by ranging them in a line, and by picking her out who projects farthest
a tergo. Nothing can be more hateful to a negro than the opposite form."
(C. Darwin 1871, II, 346)

Figure 2.3 Sarah "Saartjie" Baartman (before 1790 – 29 December 1815)
was known as the "Hottentot Venus." This cartoon, from the early part
of the nineteenth century, shows dramatically how little regard or respect
Europeans had for Africans. Reproduced with permission from Wellcome
Library, London.

After Darwin

Although, as we shall see later in this book, there are other parts of the *Descent of Man, and Selection in Relation to Sex* (to give the full title of Darwin's work on our species) which are far less time-bound and are still relevant to today's issues, there is quite a bit more of this natives-with-big-backsides stuff. One should say that, judged by the standards of the early twenty-first century, no one writing on the topic of human diversity was desperately edifying at the time. Those who favored some kind of Lamarckian causal hypothesis generally drew attention to the fact that black people live in warmer parts of the world, whereas white people live in the cooler climes. It was suggested that you have to work a lot harder the farther you are away from the Equator and that hence would acquire a bigger brain and so forth, which accounts for the obvious differences in intelligence between the races. There was, however, a certain amount of flexibility. Apart from putting the English at the top (or Germans, if you were from that country) and the Irish, or possibly the Fuegians, at the bottom, one could move races up and down according to personal preferences or prejudices. Ernst Haeckel, an ardent German evolutionist, had put the American native people (the "Red Indians") higher than the people from the Far East, partly because of mythology about the noble savage. But then he had a couple of Japanese graduate students who impressed him mightily, and in the next edition of the pertinent book the Asians had moved several steps higher.

As with evolution generally, however, in the second half of the nineteenth century, it was not really causal speculation that was the chief motivating factor. What really excited people was the possibility of finding pertinent fossils, and so the search went out for the "missing link." Other than in the American South, most people no longer wanted to put all racial differences down to the punishment of the descendants of Noah's son Ham, who sinned by seeing (and presumably making fun of) his dad when he tied one on after the stresses of the voyage. But thoughts of degeneracy died hard. Many preferred Asia over Africa (not Darwin) as the putative birthplace, mainly because they did not really want to face the awful possibility that we all might be descended from black people. Other than a few really rich men – like E. D. Cope and Otheniel Marsh, who spent fortunes fossil hunting in the American West – much of the work done generally in foreign lands on the life sciences and on the emerging social sciences was done as a sideline or hobby by people otherwise employed, as soldiers or teachers or missionaries. It is perhaps no surprise therefore that the real breakthrough was made by a Dutch army doctor working in Indonesia (Shipman 2002).

In 1891, on the island of Java, Eugene Dubois unearthed – or rather his team unearthed, because he was given not just two assistants but a squad of fifty convicts – pieces of a skull from an animal that was human-like but clearly more primitive in some sense. He called it *Pithecanthropus erectus*, using a new generic name to stress its difference from humans and its special status. More commonly it was known as "Java man." As it happens, almost at once the find was surrounded by controversy. In part it was because of the challenging nature of the discovery, compounded by the finding of a thigh bone which many thought was simply human (thus detracting from the possibility that something less than human nevertheless walked upright). In part, this was because of Dubois's own prickly nature, leading him to sequester the findings for many years. Today, it is recognized as a truly important find, although reclassified as a member of our own genus as *Homo erectus*. Contrary to a myth much favored by Creationists, Dubois did not end his days denying that Java man was in any way connected to humans. What he did start to do was emphasize its ape-like nature, thus highlighting how it was not truly human.

We now come to one of the greatest frauds in the history of science: Piltdown man (F. Spencer 1990). In 1912, a British amateur archeologist, Charles Dawson, claimed to have found, in England, south of London, frag-ments of a skull that was remarkable for being an amalgam of a human-sized brain and an ape-like jaw and teeth. There were those who doubted its authenticity from the start, but general opinion was that a real link had been found – one clearly suggesting that crucial episodes of human evolution had occurred in Europe and that most probably the brain had exploded upward in size before the rest of the ape body was transformed into something human-like. Further fragments were found a year or two later, but no more after that – suggestively, no more after Dawson died in 1916. Piltdown man, named after the village where it was found, held sway for thirty or more years, and undoubtedly bent the course of our understanding of human evolution. In the 1920s, Raymond Dart in South Africa described a newly found fossil from that part of the continent, the brain of a small creature which undoubtedly walked upright (Lewin 1987, 1989). Taung baby, as it is now called, the first specimen of what is now considered an ancestral genus separate from ours (*Australopithecus*), was dismissed as insignificant at least in large part because it directly contradicted Piltdown man, suggesting that it was bipedalism that came first with a large brain following after. Clearly also a factor was that this would seem to bring us full circle and back to Charles Darwin's belief that Africa was the cradle of humankind.

Finally, in the early 1950s, Piltdown was shown to be completely fraudulent, the skull of a modern human and the suitably stained jaw of an orangutan, with a few chimpanzee teeth thrown in for good measure. Even to this day, there is debate about the perpetrator. One of the British Museum investigators of the find was Martin Hinton, and a decade or so ago people found the requisite staining chemicals in a long-neglected trunk owned by him (Gee 1996). But it is unlikely that he acted alone, and suspicion still points also to Dawson, who is known to have had a record of dishonest behavior in the archeological field. Perhaps more interesting than who did it is the way in which so many ignored so many suspicious signs simply because they had prior prejudices about the course of human prehistory. One of the later-discovered fragments is carved in the form of a cricket bat. Viewing it with hindsight, as I did a few years ago in the Natural History Museum in London, it seems incredible that anyone was taken in by the hoax.

Back on track, the middle of the twentieth century saw increasing numbers of fossil finds, confirming that Africa was indeed the birthplace of *Homo sapiens* – most probably central East Africa, Ethiopia, Kenya, and near countries. Much credit goes to the indefatigable Leakey family: Louis, his (second) wife Mary, and somewhat later their son Richard. Finally, in 1973 the American paleoanthropologist (as students of the human fossil record are known) Donald Johanson and others working in Ethiopia made the greatest discovery of them all, the Platonic form of missing link (Johanson and Edey 1981). Lucy, as she was called after the Beatles' song, now classified as *Australopithecus afarensis* and about 3.25 million years old, walked upright (although probably not quite as well as we humans), yet she had a chimpanzee-size brain. Note that it was a chimpanzee-size brain (400 cc as opposed to our 1,200 cc brain), but it does not follow that it was a chimpanzee brain. She was our ancestor or near ancestor.

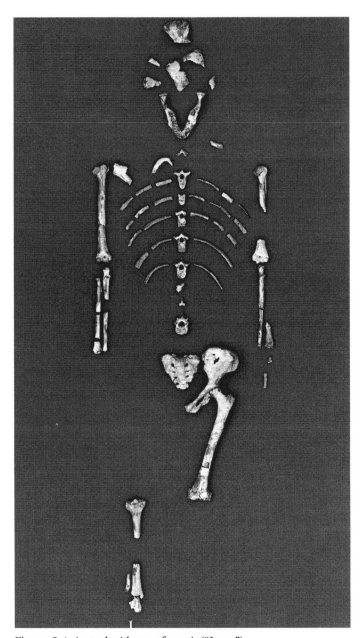

Figure 2.4 *Australopithecus afarensis* ("Lucy").

More recently there have been other discoveries, some of older specimens and some of younger specimens. Among the older, particularly
impressive are specimens of a genus, *Ardepithecus*, that lived just before
Australopithecus (White *et al.* 2009). *Ardepithecus ramidus* lived about a million

years before Lucy, and was even more ape-like in respects, particularly inasmuch as although she (the best-known specimen is female) could walk upright, it is clear that she was far better adapted for tree climbing than Lucy, who in turn was probably better than us. There are earlier specimens than *Ardepithecus*, although (perhaps as a function of the fact that they were jungle dwellers) very little direct fossil evidence of the great apes (chimpanzees, gorillas, and orangutans). Among the younger specimens, truly astonishing is the discovery early in the new millennium of a race of very small human-like beings – inevitably nicknamed "hobbits" – living on the island of Flores in the Indonesian Archipelago, dating from less than 20,000 years ago (Brown *et al.* 2004). There has been much robust discussion about their authenticity, with some still arguing that they are diseased or otherwise handicapped members of our human species, but general opinion does seem to be that they are genuine and that they may possibly be more distant relatives than was at first assumed.

Misconceptions

Before we turn and talk briefly about causes, a number of misconceptions should be noted, especially since critics of evolution often seize on and exploit them. First, although the fossil evidence is tremendously important, it is not the only clue to uncovering the human past. For the past fifty years, molecular biology has been a major part in the toolkit of those who search for human origins. We have encountered already the notion of genetic drift and seen that although it clearly does occur, it is probably not a significant factor in the general physical and behavioral features of organisms. Selection will swamp its effects before they can be that lasting. As noted also, at the molecular level, the story is probably different. There it is clear that many elements do little or nothing to make the whole organism, they are "junk," and there are lots of redundancies and duplications. Selection simply can have no effect on their ratios, and so their numbers drift aimlessly through time. This is the "neutral theory" of evolution already mentioned. Seizing on this, evolutionists have exploited the subsequent drift to make a "molecular clock," arguing (reasonably) that since things will average out, one can use molecular differences in groups to calibrate time spans – essentially, twice the difference, twice the time. Of course, one needs some external markers to get absolute times, but

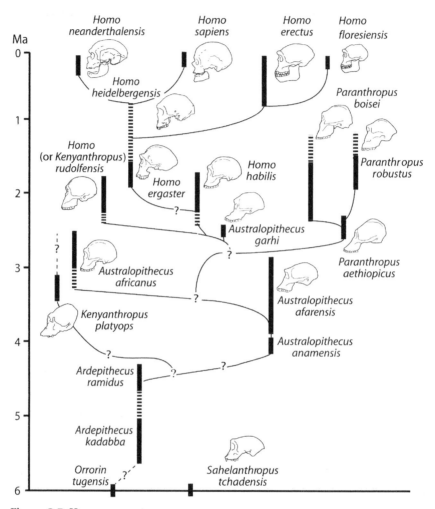

Figure 2.5 Human ancestry.

with these in hand, one can make very sophisticated estimates of the dates of significant evolutionary events, especially of when today's groups may have split apart (Ayala 2009).

And the results in some cases have been very surprising (McHenry 2009). None more so than in the human case. Around 1960, pretty much everyone in the field – judging intuitively from the differences between humans and the great apes, although no doubt also guided by prior beliefs about our distinctive nature – put the break between us and the apes somewhere around 15 million years ago, more or less. Then the molecular biologists brought the date down dramatically to about 5 million years ago. And in that region it has stuck, despite the initial, almost frenzied opposition of the students of the fossil record. It is a conclusion reaffirmed again and again, not the least by the fact that from a (molecular) genetic point of view, humans and the chimpanzees and gorillas really are very similar. Indeed, as is well known, it is now thought that humans and chimpanzees (there are two species) are closer together – their split occurred more recently – than either of these species and gorillas – that split was more ancient.

Second, note that if the tree of life is fairly bushy, with lots of speciation, although every new find is hailed as that crucial ancestor, it is improbable that any find is indeed the actual ancestor. Most fossils are likely to be a bit off on the side. There is nothing mysterious about this, or particularly threatening. It certainly does not disprove the importance of the fossils. Suppose for instance a new adaptation A turns up – like walking upright – and this occurs let us say 4 million years ago. Then in the next million years you might quite possibly get this group splitting into three different species, let us say x, y, and z. Why? Usually, because small groups get geographically isolated from the main group and then evolve away. Perhaps (possibly like the hobbit) they sail successfully over dangerous seas and find themselves on an island, far away from home and unlikely ever again to see or rejoin the parent group. However, the main inference still stands. It is pretty clear that humans today come from this overall group somehow, because we uniquely have adaptation A. Suppose now, from the overall group, we find fossils of group y. Unless there is some other evidence helping us decide, we ourselves could have easily come from x or z, as from y. Hence it is more than possible that the fossil discovery is not of the exact ancestor. Equally, however, discovering fossils from y is throwing light on the nature of x and z also. So it is hardly the case that the fossil record is irrelevant. Of course there are issues here about whether A might have been developed independently on several occasions, but there are techniques to cut down on this likelihood. Mainly there is the obvious point that if the adaptation is fairly sophisticated and we have it exactly, it is unlikely that it evolved again in exactly the same fashion.

A third point is about the nature of species. We saw in the last chapter that Darwin prided himself on finding a reason to think that Linnaean classification could be natural, meaning objective, and not just a matter of convention or convenience. This was because the classification represented ancestry. However, you might think that this cannot be entirely true, because after all the whole point of the Darwinian scenario is that life is fluid, with one form sliding gradually into another. Surely at some point any divisions that we make have to be arbitrary? The answer to this question is "yes" and "no"! No, in the sense that if we are dealing with a tree (or net) of life, we may have trouble finding it, but its existence and nature are independent of us. So that is not arbitrary. Yes, in the sense that exactly how we try to cut things up might be our decision, whether done because that is what we want to do or because of ignorance.

It is generally agreed – Darwin agreed much of the time – that taxa at one particular category level, the species, are objective in a way that other taxa at other levels tend not to be. What is a genus (the level above species) for instance? Why should we say that Lucy is in the genus *Australopithecus* and not the genus *Homo*? Obviously in main part because we want to say that the australopithecines are quite different from the members of *Homo*. But if one of the later australopithecine specimens were shoved in with us or if one of our earlier specimens were shoved in with them, it would not make a great deal of difference. However, in the case of species, like *Drosophila melanogaster* (a species of fruit fly) or *Homo sapiens*, it is thought that there is an element of objectivity. Suppose that the little hobbit, *Homo floresiensis*, were still around today. (It is thought to have gone extinct less than 20,000 years ago.) We have a criterion (known as the "biological species concept") as to whether we are the same species as them or not, namely can we interbreed with them – not just have sexual intercourse but produce fertile offspring (Mayr 1942). If we could, then we are one species. If we couldn't, then we are two species. This is true, notwithstanding the fact that, as Darwin pointed out, if one group splits into two, you expect that there are going to be borderline cases.

The biological problem is about interbreeding and how it fails (Coyne and Orr 2004). The philosophical problem is about why we should think that reproductive isolation (as the barrier between species is called) should be thought to confer objectivity or naturalness. There has been much discussion of this issue. Some think that it is because an interbreeding group (and only an interbreeding group) takes on the characteristics of an integrated whole, like an organism (Ghiselin 1974; Hull 1978). This is known as the "species as individuals" thesis. Others find this solution unconvincing. They argue that you do not find the integration between members of a species that you find between parts of the body. (This debate in turn gets caught up with the "levels of selection" debate, about whether selection favors only individuals or sometimes groups as well. If selection always favors individuals, then talk of integration seems unwarranted.) These critics often prefer a more philosophical approach which owes something to Whewell's thinking about the virtues of a consilience. (Whewell himself saw the connection.) They argue that reproductive isolation is connected with other features of the organisms, for instance those to do with physical looks and behavior. Humans just don't look like hobbits, nor do we do the same things as they did. So the objectivity comes because all sorts of different features focus in on the same collection, just as all sorts of different areas of science focus in on the same uniting hypothesis (Ruse 1987a).

Either way, obviously, even if you agree that calling something *Australopithecus afarensis* or *Homo sapiens* is to say something about objective grouping, you still have an element of arbitrariness. *Homo habilis* led to *H. erectus*, led to *H. sapiens*. Unless punctuated equilibrium is true, and most people don't think it is true in any extreme form, this means that you are going to have border regions with organisms between *H. habilis* and *H. erectus* and then between *H. erectus* and *H. sapiens*. It is facts like this that can make it so difficult to decide whether a group like the Neandertals should be considered full members of our species or not – and often in the end there is no real answer. Fortunately however, often what we cannot do nature does for us. We only have occasional fossils from the record and what we have are reasonably distinct. In other words, as a practical matter we use the gaps in the fossil record to break species apart, although we know that in truth we are doing this only by pretending that the absent organisms really never existed!

Why humans?

Turn now to the full picture and to causes. The dinosaurs (with the exception of those who stayed on as birds) became extinct about 65 million years ago. Mammals had been around for many years before that – probably as long as the dinosaurs themselves, going back to 200 million years ago – but it was only with the demise of the giant reptiles that mammals literally came out of the nocturnal undergrowth and started to flourish. Primates appeared soon after that, and then some 20 million years ago or so came the break between the monkeys and the apes. Jumping ahead rapidly, and recognizing that by focusing on humans (because we are human!) it does not necessarily follow that our evolution is the most interesting aspect of all of evolution – we shall be addressing this issue directly in a future chapter – Why did human evolution occur? Why did we get up on our hind feet? Why did our brains get so much bigger? Let us start with the first, and as we now know earlier, event.

The great apes can walk, but to do so they tend to use the knuckles of their fists to steady themselves (Relethford 2008). What if anything does this tell us? Were we hominids (as the great ape and human combined group is called) all up on our hind feet, were we all bipedal, and then the apes slouched down and became knuckle walkers? Did we all have something different, say some sort of climbing ability, and then did we hominins (as the humans and their post-split ancestors are called) become bipedal and the great apes knuckle walkers? Or were we all knuckle walkers and then we humans became bipedal? As far as the first hypothesis is concerned, you need a reason to become knuckle walkers. Not impossible of course, especially if you firmly squelch human-directed feelings that walking upright is somehow "nicer" than going through life bent over and using your forelimbs to walk. But in any case, probably not the first and second hypotheses, because these would imply that gorillas and chimps gained knuckle walking independently – not impossible, but not the simplest solution. (Always a major point for scientists.) So this really suggests that we all walked on our knuckles and then for some reason the hominins started to lose the practice. There is indeed some fossil evidence that the early hominins do show a greater capacity for knuckle walking than do later hominins.

But why? Thanks to human encroachment on their territory, the great apes today are in dire straits in the wild. But in the past, why should the apes be less well off than we humans? They can get around pretty well using their knuckles, and they are obviously far better than we humans when it comes to life in the trees. Why drop down onto the ground, especially when you obviously put yourself at risk from predators? A popular hypothesis used to be that a major reason for bipedalism was prolonged African droughts. The jungles started to dry up, and grasslands, savannah, came in their stead. Hominins were forced out into the open. Perhaps it was we who failed, because the apes were better at jungle life than we, or perhaps they kicked us out and we had to make do elsewhere. Out on the grasslands, with all of those proto-lions and tigers, you needed to be upright to look around you and spot the dangers. Selection worked flat out to have us strutting around like a bunch of grenadiers.

Thomas Henry Huxley once said that Herbert Spencer's idea of a tragedy was a beautiful hypothesis destroyed by a nasty little fact, and I am afraid that we have a tragedy in the making here. The ever-increasing fossil record, especially the newly reconstructed *Ardepithecus*, shows that our ancestors started to come down out of the trees and walk upright even though their habitats were still comparatively wooded (McHenry 2009). They started the route to bipedalism before they were pushed out onto the grasslands (if indeed they were so pushed). So "got-up-and-looked" will not work. Neither, at least in its original form, will "needed-hands-free-for-tools." This hypothesis basically sees big brains and bipedalism coming together, as our ancestors started to rise up and at the same time, with hands free, started to use tools and manipulate the environment in that way. But we know now that bipedalism came before the increased brain size that seems required for sophisticated tool use. This is not to deny that there may be something to a lesser form of the hypothesis, with hominins using crude tools. After all, chimpanzees do that much.

Other hypotheses are also controversial. It has been suggested that we may have stood upright in order to carry children and food, especially when foraging (Lovejoy 1981). Part of the problem here is that this model in its original form posited monogamy, with one male and one female working together and thus without the immediate help of others. The sexual dimorphism of australopithecines suggests, however, that our ancestors were polygynous (one male, several females). This may incidentally all sound a little bit like Darwin's thinking in *Descent*, and of course in a way it is. The point today is that it would be emphasized that this is all a two-way thing. It might well be advantageous for a female to ally herself with a powerful male, even though she has to share, rather than have unique access to a less powerful and successful male. (In fairness to Darwin he does allow for the process of female choice, but in discussing humans in the *Descent* this is usually forgotten or ignored. It is he-men all of the way.)

Other hypotheses include the suggestion that being upright is a better way to avoid heat stress – less sun strikes a vertical body than one spread out as with a knuckle walker – and that although bipedalism may not lead to the fastest animals (knuckle walkers are better fast runners), at moderate walking speeds bipedalism is a much more efficient way to travel (Wheeler 1984). Yet although both of these hypotheses are based on solid evidence – and obviously are not mutually exclusive – if we then all lived among the trees, neither seems to account for the beginnings of bipedalism. The sun is not a big issue and neither is walking. However, if one supposes that perhaps bipedalism started with things like foraging on the ground – chimpanzees stand upright when they do this – and that then, as the jungles started to vanish, there was need of adaptations to move from one clump of trees to the next, one can see how many or all of these hypotheses could kick into play as bipedalism became more and more efficient (Larsen 2008).

Whatever the exact cause of bipedalism, and remember that what triggered it is not necessarily why it was picked up and perfected, the main point is that at some point around the 5-million-year mark hominins were starting to walk, their hands were freed up for other tasks, and also things were indeed drying up and forcing us out of the jungles and onto the grasslands. Certainly one need is to find new sources of food as an alternative for all of those juicy fruits to be found hanging from trees. And there was one obvious source, namely the big chunks of protein to be found out on the savannah – that is to say, other mammals. Chimpanzees and other apes will eat protein if they can get it – little monkeys and so forth – so there was probably no real switch here, rather a move to increased reliance. This was a two-way phenomenon. On the one hand, there was need to get the protein. On the other hand, there were the consequences of the protein. As far as getting the protein, note that outside L'il Abner cartoons animals do not lie down and die of joy at the thought of being eaten. You or someone or -thing has got to catch it. It is thought that this was a major factor triggering tool use – crude implements to kill the prey and to butcher it. Another major factor was the need to cooperate. Hominins are not that fast and certainly not that strong, comparatively. You need to plan and to work together to get your prey. Or, as the evidence of the bones suggests, to get the prey that others have caught and killed. Rather than hunters, there is good evidence that our ancestors were scavengers, taking the remains that had been brought down by natural predators, lions and so forth (Isaac 1983). All of this obviously puts a major pressure on increased skills and intelligence, which in turn demands bigger brains. Here the second part of the equation kicks in. Brains are expensive to build and to maintain. They require lots of energy. Animal meat is the best of all fuels, far better than grass or fruit or whatever. In the twenty-first century you can live the life of a vegetarian or even a vegan, but you should not kid yourself that it is natural, if you mean that it is what nature intends and what our ancestors followed before there was civilization.

Homo sapiens

Rather less than 2 million years ago, bigger, brainier animals had emerged – our genus, *Homo*. With them come ever-better tools and the use of fire, although there is discussion about the exact date when this latter was truly brought under our control as opposed to used on occasion. At the same time, the ability to talk was starting to emerge in a big way. I don't think anyone today wants to deny that biology plays a big part in language acquisition and use. In the 1950s, Noam Chomsky (1957) showed convincingly that language is not just something purely cultural but that all languages, from Japanese to English, share certain innate deep structures – a kind of biological ground plan on which everything is based. As it happens, Chomsky is not particularly keen on Darwinian selection, but his students and followers like Steven Pinker (1994) have shown in detail how the innate hypothesis does lend itself to Darwinian understanding. It is hardly the case that one has to strive very hard to make the case for the great adaptive significance of language.

Complementing this grasp of the software, as it were, the understanding of the hardware of language ability has grown greatly (Holden 2004). There is some fossil evidence of the actual physical apparatus needed to speak. The dropping of the larynx, for instance, is something which distinguishes us from the apes. (Something else will be mentioned very shortly.) There is also evidence of the development starting about 2 million years ago of parts of the brain that are used in speech, specifically Broca's area and Wernicke's area. (No one has the actual brains, obviously, but from casts of the insides of skulls, one can infer what was there and if it was increasing.) Relatedly, we are now starting to identify some key

genes involved in language acquisition and use and can show that they are
under the force of natural selection.

With the appearance of the genus *Homo*, and then on to our own species
Homo sapiens (generally put at about half a million years ago), two major
hypotheses now dominate the discussion. Was it the case that the real
evolutionary action generally or always took place in Africa? Was this generally followed by successive migrations out of Africa and into the rest of
the world (Stringer 2002, 2003)? Or was it the case that the migration took
place relatively early, and that then parallel lines of hominins evolved up
to the present (Wolpoff and Caspari 1997)? If the latter, then probably (even
necessarily) there was some gene exchange between the lines, because it is
highly improbable that the lines could have turned into modern humans
independently.

General sentiment among paleoanthropologists is that the out-of-Africa
hypothesis is the more likely and gets the nod over the multiregional continuity hypothesis. The fossil evidence bears this out, as does other information, for instance the lack of genetic variation in humans as opposed
to other animal species, rather suggesting that not too far in our past (as
recent as 60 thousand years ago) modern humans went through a bottleneck with numbers much reduced, as one might expect if we are all (or
most of us) descended from a relatively small group making its way out
of the African continent. However, matters are by no means settled absolutely, and there is strong evidence of early humans at least up to a hundred thousand years ago in the Middle East (where Israel is now). This
information is especially pertinent when it comes to a group that precedes
ours and also spread itself around, the Neandertals.

When we last met the Neandertals, although Thomas Henry Huxley
was prepared to include them in our species, they were getting a rather
bad press – short, squat, hairy, brutish, and not very bright. Certainly not
the type you would want to meet in a dark passageway at night. Recently,
however, the Neandertals have been getting quite a good press, starting
with the fact that if anything their brains were rather larger than ours!
Of course, you cannot really exploit this if you cannot communicate efficiently, and for a while a popular hypothesis was that the Neandertals like
the great apes seem to lack the ability to talk properly (Lieberman 1984).
But this suggestion has now come crashing down with the discovery of a
Neandertal hyoid, a bone that is found in the throat and has essentially

the function of enabling speech (Arensburg *et al.* 1989). No one is suggesting that there was a Neandertal Shakespeare – in fact, general opinion (based on the rise and development of culture) is that it was not until about 50,000 years ago that true speech emerged, and that it was probably some kind of primitive click language, found almost exclusively in Africa. But the Neandertals were not dumb brutes.

It is true that the Neandertals did tend to be shorter and stockier than humans, but this is almost certainly not a fault but an adaptation for cold climates – it is a matter of surface area to body mass and of conserving heat (Relethford 2008). The same is true of Neandertal noses – they were large, if not as monstrous as Pinocchio's after telling fibs, but again that is what you want in cold weather (for warming the air as it goes down to the lungs). It turns out that they were pretty good hunters, if one judges from the remains of animals in their dwelling places, as well as by more sophisticated methods such as those involving microstudies of teeth showing that their diets were as rich and regular as those of more modern types of humans. There is even evidence that they had rituals, such as surround the burial of the dead (Stringer and Andrews 2005).

What about the $64,000 question? Are we Neandertals? At least, does some Neandertal blood flow in our veins? It is thought that humans and Neandertals diverged somewhere between a quarter and half a million years ago, and there is evidence of the Neandertals very early in Europe. They became extinct about 30,000 years ago. So there was certainly opportunity for interbreeding, because sometimes modern humans and Neandertals lived in the same areas. It is now possible to extract DNA from Neandertal bones (Gibbons 2010; Krings *et al.* 1997). Initial studies made on the DNA from mitochondria (these are not part of the nucleus of the cell but the power plants that drive the cell) suggested that Neandertals really are quite different, enough even to be considered a separate species. However, the recent sequencing of nuclear DNA has offered a very different story. There is incontrovertible evidence of some interbreeding between Neandertals and the early human settlers in Europe from Africa. It seems that somewhere between 1 and 4 percent of human DNA came from the Neandertals! Not a huge amount, admittedly, but not negligible either. What is fascinating is that it seems that it was the early settlers who did the interbreeding, and that later coexisting groups stayed separate. Could this have been a function of cultural differences? What is even more fascinating, although perhaps to be expected, is that there was no reverse gene flow back to Africa. It is Europeans and Asians and others who have Neandertal genes, not Africans. There is an irony in who is related to cavemen and who is not, something that will not be lost on us when later in the book we come to discussions of race.

No one is looking for this kind of relationship with *Homo floresiensis* (the hobbit). It is most probably something altogether different – it could even be that it broke from the human line as long ago as the australopithecines. If this proves to be so, while it would be very exciting, and would surely demand some serious rethinking by those who think that humans are really special, it would not worry paleoanthropologists a great deal. Being a very exciting discovery does not make it theory wrecking. We have seen that the whole picture of evolution is one that makes branching absolutely central, and this would be just one more instance. Moreover, there is already some good theory to deal with some of the hobbit's most distinctive features. It is well known that being isolated on islands tends to lead to extremes in body morphology. You get gigantism, as with the huge birds in New Zealand that were wiped out by the arrival of the Maoris in the past thousand years. There were no mammals occupying the top predator niches and so the birds stepped up to the plate to do the job. You also get dwarfism, a function of the lack of food in isolated areas. With no competition to go otherwise, organisms often shrink right down and live very comfortably. On Flores, there were dwarf elephants to match the hominins.

I should add in concluding this section that this whole question of recent species (or subspecies) is a very fast-moving area of inquiry and one is out of date almost as soon as one has put pen to paper (or the modern equivalent). The most recent finding is that the Neandertals were not one uniform group but were split into the more western traditional Neandertals and an eastern group, now labeled the Denisovans. The Neandertals and modern humans split about 500,000 years ago and then the Denisovans split from the Neandertals about 100,000 years later. It is thought that they may have ranged from Siberia to Southern Asia, and there is a good chunk of Denisovan DNA in the genomes of today's inhabitants of New Guinea (Reich *et al.* 2010). Again, let me stress that exciting though discoveries like this (and undoubtedly more to come) may be, they round out the picture. They do not challenge the basics.

Culture and consciousness

Drawing to an end, we move down to the present. Agriculture is a big innovation here (Larsen 2008). Somewhere about 10,000 years ago, humans stopped going out and hunting and foraging and started to grow crops themselves and to domesticate animals, beasts of burden and those that could be eaten. No one is quite sure exactly how or why this happened, although it seems clearly to have been related to the fact that climatic conditions calmed down from irregular periods including ice ages to a fairly comfortable plateau of warm and wet, and stable, conditions – just what the farmer needs. At first it was all a bit haphazard, with the gathering of grains at certain times of year and older practices prevailing for the rest of the time, but then increasingly people managed to cultivate crops and to supply needs on a regular basis. Human numbers started to grow, too, no doubt in a kind of feedback system, for as agriculture led to more regular supplies of food, fewer would die of starvation, but this in turn put pressure on more, and more dependable, supplies of food. This led to the possibilities of larger and larger gatherings of people, and the modern world was finally in sight.

There is no need to go into more detail here. Throughout this book we shall be supplementing the exposition in this chapter. Note, however, that although evolutionists can tell us much about human origins, on some issues – issues that are certainly of interest to those of us with philosophical interests – evolutionary biology is going to say little or perhaps even nothing. Most obviously there is the issue of consciousness. The evolutionist is not exactly scared of consciousness (Pinker 1997). Pretty clearly it is something that we use to guide us through life. I don't suppose a bee is particularly conscious. It does everything by instinct. It is a living machine. We are conscious, and we use our consciousness to make decisions and so forth. Some paleoanthropologists – or more commonly, archeologists, being the people who deal more directly with the civilized side to the human past – have come up with quite elaborate scenarios explaining how our thinking developed over the eons. A popular, although not uncontested belief is that the human brain is built on the modular principle, with different parts doing different tasks and the whole connected up rather like a Swiss army knife (pro: Fodor 1983; con: Sterelny 2003). Utilizing this idea, British archeologist Steven Mithen (1996) has put together an evolutionary picture that ties in increased use and sophistication of tools with the growth of brain size, something he emphasizes does not come altogether gradually but rather in leaps and bounds with periods of stability. We start with a kind of general intelligence, of the sort possessed by the apes, that functions well for social situations and for navigating the environment. Then (with the arrival of *Homo habilis*) we get a spurt that leads to modules for technical ability and so forth, but nothing hugely innovative and probably not a great deal of technical information being transmitted. As we move to *Homo sapiens*, we get the development of more modules and eventually about 50,000 years ago things really start to take off. (And that is the point when the Neandertals really do get left behind.)

But why consciousness? Even if all of this is true, why are we not simply automata? Thomas Henry Huxley (1874) used to think that consciousness simply sits on top of the brain, working in a purely mechanical method and doing everything that is necessary. Known as epiphenomenalism, more informally it has been characterized as the "whistle-on-the-locomotive" position. I don't think many people find this very convincing. It basically says that the mind has no adaptive function, although remember that for Huxley this was less of a worry than it would be for one who took natural selection seriously. It does seem that the mind influences the brain and the physical and is not just redundant fluff. In the words of William James: "It is to my mind quite inconceivable that consciousness should have *nothing to do* with a business which it so faithfully attends. And the question, 'What has it to do?' is one which psychology has no right to 'surmount,' for it is her plain duty to consider it" (James 1880a, I, 138).

But why bother with consciousness at all? Why not simply take it out of the equation and have the brain do everything? Or if the brain cannot do everything, why or how does consciousness arise from a purely material thing like the brain? I think here we start to come to the limits of what an evolutionary theory can do or can even be expected to do. The DNA molecule is obviously a very important part of the evolutionary story today, but no one expects Darwinian theory to explain the molecule – how for instance it can form a double helix. This is part of the chemistry of the world. Natural selection takes hold of the properties and works with them. If the molecule did not pair up as it did, at least in this respect it would be of no use to natural selection. It is the same with the mind. As is well known, there are various different philosophies of mind (Ruse 2010).

These range from the belief that the mind and the body are one and identical (monism) all the way to the belief that the mind and the body are two different substances entirely separate (dualism). Along the way there are those who think that there really is no mind–body problem, those who think that there may be a problem but that a bit more science will put things right, and those who think that there is a problem and that we will never solve it. This last group, thinking that it is all a mystery, borrow the name of a pop group and refer to themselves as the "new mysterians" (McGinn 2000)! For the moment, perhaps it will be best to leave things at that. No one is claiming (at least not qua science) that the mind is some supernatural object, best left to the theologians. Whether it is material or not, no one is saying that it is not a natural phenomenon – a part of the physical world, in the general sense (just as electrons with their odd properties are part of the physical world). No one is saying that the coming of the mind and the use of the mind are not problems for the evolutionist, whether good answers can be given or not. But the actual existence of the mind – the actual existence of sentience – and its relationship to the material might perhaps be beyond evolutionary understanding and no business of the evolutionist anyway.

3   Real science? Good science?

I want now to look at the related (philosophical) questions of whether with Darwinian evolutionary theory we are looking at genuine science and, if so, at good science. Without feeling obliged to keep the two parts of the discussion totally separate, I start with the general picture and then turn more specifically to the human picture.

Darwinian theory as science (laws)

The Creationists deny that we are dealing with genuine science. They say that Darwinism is "just a theory not a fact," but of course they think it is at best a false theory and truly not even a real theory at all. However, as many have pointed out, there is an ambiguity at work here. There are two senses of the word "theory." In the one sense, we are talking about a body of laws, put together to explain some part of experience. Plate tectonics is a theory in this sense. In another sense, we are talking about an iffy hypothesis, as in: "I have a theory about Kennedy's assassination." Plate tectonics is certainly not a theory in this sense. It is a true claim about the world. Continents do rest on big flat areas of rock that slide around the globe.

Now what about Darwinism? Clearly in the sense of a body of laws, meaning a body of claims about universal patterns or generalities in nature, Darwinism is a theory. Sometimes when people are talking about laws and evolution, they refer to generalities that were popular early in the history of the theory. Bergmann's law states that organisms tend to get bigger as they get farther from the Equator. Cope's law states that in lineages of organisms, individual members tend to get bigger over time. In fact, generally speaking, these laws (or "rules" as they are often called) are not really what is at issue today. For a start, they are not truly that evolutionary. Bergmann came up with his rule in 1847. And although they certainly have some validity and can be based on evolutionary principles – Bergmann's rule can be interpreted in terms of body mass versus overall surface area, and of how a bigger organism is more efficient from a heat-loss perspective (note that *Homo sapiens* is a lot bigger than *Australopithecus afarensis*) – there are just too many exceptions to instill great confidence. *Homo floresiensis*, for instance, almost certainly had ancestors that were larger that it was. In any case, the laws or rules are not really that basic to the theory. We want some evidence that the heart of the theory appeals to laws. But this seems to be no great problem. In the first chapter, we saw how Darwin in the *Origin* appealed to laws: organisms have a tendency to multiply geometrically, natural populations have variations, and so forth. Then remember, in today's version, that we start with genetics and only later introduce selection. We start specifically with the Hardy–Weinberg law, which acts like Newton's first law of motion. They are both equilibrium laws: if nothing happens, then nothing happens! Against this background of stability, we can start to introduce disruptive factors, like mutation, migration, and so forth. Especially natural selection.

So no problems here. Or are there? Are we not perhaps being a little bit quick? Darwin's claims may be better than Bergmann's rule, and the Hardy–Weinberg law is certainly better, but are they good enough? Darwin himself recognized that there are going to be exceptions. "Although some species may be now increasing, more or less rapidly, in numbers, all cannot do so, for the world would not hold them" (C. Darwin 1859, 63–64). It is all somewhat a case of damned if you do, damned if you don't. Laws are supposed to be universal, so you have got to do something about exceptions. Laws are also supposed to be true, and not just true but in some sense necessary (Hempel 1966). They are not just contingent generalities (such as everyone in my undergrad philosophy of science class is a native-born American), but things that have got to be true (such as all bodies attract each other with a force inversely proportional to the square of the distance between them). A law with exceptions is false and hence hardly true, let alone necessary. One way you can deal with exceptions is by adding on *ceteris paribus* clauses – all other things being equal (which they rarely are) – but then you run the danger of making the laws simply true by fiat. That is to say, they become analytic or tautological, true virtually by definition. As it happens, a charge along these lines is often made against Darwinism, specifically with respect to the central mechanism of natural selection. The alternative wording (coming from the British scientist-philosopher Herbert Spencer and adopted by Darwin in later editions of the *Origin*) is the "survival of the fittest." But who are the fittest, ask the critics? Obviously they are those that survive. Hence, natural selection reduces to the empty statement that those that survive are those that survive. This is true, but not very informative about the real world. (This criticism is a favorite of those who deny the truth of evolution. See, for instance, Johnson (1991). It is also made by some who accept evolution but who have qualms about Darwin's theory. See for instance Popper (1974). Rosenberg and McShea (2008) have a full if somewhat skeptical discussion of laws and theories in biology. Ruse (1973) is very dated in some ways, but it covers a lot of the basic material.)

Digging out from under this barrage of complaints, at a certain level you have to take a realist or pragmatic attitude to the way real science is done – in physics and chemistry as well as biology. I suppose that most of us (those who take science seriously) think that ultimately there are laws of nature, without exceptions, that govern the universe. These probably hold down at the molecular or sub-molecular level. But then, as you start to deal with things at larger and larger levels, exceptions and qualifications appear all of the time. Take Galileo's law about bodies falling to the ground at constant acceleration. Boomerangs do anything but! Yet no one loses any sleep about this. You have to put in qualifications about the air pressure and so forth. It is the same with Boyle's law. It does not hold at extreme temperatures and pressures. Again, qualifications are put on it, and in a way the exceptions are illuminating, because they lead to investigations of the extremes, yielding (in this case) Van der Waal's equations.

Darwinians point out that they are no different. One expects that normally, that is except when restrictions are lifted, population numbers are going to explode. Indeed, when they do not explode, that is when it gets interesting. It is the same with variation in populations. Of course, artificially, you can aim for a population with no variation – that is what medical researchers do with their pigs and mice. But generally you expect variation. And again, when you don't have variation, that is when it gets interesting. In fact, you have a nice example of this in the human case. You do not get the genetic variation in humankind that you might expect, and that you get in the great apes, for instance. This does not mean that you give up on the laws of nature. Rather, it means that you look for causes making for the exception. And as we have seen, such causes have been postulated – they are the mainstay of the out-of-Africa hypothesis, suggesting that the modern human species went through a bottleneck (thus reducing human variation) not that long ago.

What about the charge that natural selection is a truism? Well, at one level, we know at once it cannot be completely true. Apart from anything else, natural selection depends on there being a differential reproduction – some will survive and reproduce and some will not. This is true, but it is certainly not tautologically true. You cannot decide on it by simply thinking. You have to go out into the world and check. Moreover, take genetic *Solution to* drift. Many are not very keen on it, but no one says it is a contradictory *the tautology* idea, which it would be if natural selection were a tautology. Sometimes *problem?* the fittest don't survive and reproduce. In fact, once you see what the criticism is missing, you can see why it fails, but also why it is so seductive a charge. The Darwinian's claim is not just that there is a differential reproduction, but also that certain features aid or help organisms in the struggle – they are what help them to survive and reproduce. These are labeled the fittest, although note that it is an average that is talked about. No one denies that sometimes through chance or bad luck the very good loses and the very bad succeeds. But on average this does not happen. It is also thought that things that help in one situation will help in similar situations, too. Being white in the snow helps the bear. It also helps the fox. Of course the trouble is that of deciding exactly what are similar situations. And note that we have a kind of inductive assumption built in here. What *The problem* holds in one part or time of the world, holds in other parts or times of the *of induction* world. The philosopher Karl Popper always loathed inductive reasoning and perhaps this goes part way to explaining why he was always iffy about Darwinism and brought up the selection-is-a-tautology charge.

You might think that with the coming of Mendelian genetics, followed by molecular genetics, much of this discussion is otiose. Have we not now moved on to a level of science with firmly established laws, a degree of sophistication above the science of the *Origin*? To a certain extent, this is true. There is no question that modern evolutionary biology is much more precise, thanks in large part to the extensive use of mathematics, than was the science produced by Darwin. However, even here there are critics and naysayers. Richard Dawkins (1983) has suggested that throughout the universe, if there is life then there will be natural selection. This may well be true. But what about the principles of genetics? Dare we think that on Andromeda Mendel's laws hold? There are those who wonder about how well they hold here on Earth (Beatty 1981). Even at the molecular level, can we expect a DNA world? RNA seems to be able to do a lot of the work – most researchers now think that early life was all RNA – so must we think that elsewhere in the universe this or other molecules are the key to life? In any case, charge the critics, when it comes to humans the whole question of evolutionary laws becomes dubious, if only because laws have to be general and you should not refer to just one body of organisms in one time and place.

At one level, these really are philosophers' worries. Until we find life on Andromeda or elsewhere, we might just as well go on doing the evolutionary biology that works down here on Earth. At another level, we go back to the all-other-things-being-equal ploy. If it did turn out that Mendelian/molecular genetics was the exception rather than the rule, we would simply have to restrict it to the kinds of cases where it does work. After all, we are already doing this when we deal with organisms that do not have sexuality. As far as the reference to individuals is concerned, seeming to threaten the generality that characterizes laws, note that we already do this in physics. Kepler's laws apply to our planetary system. Note also that the reason that we feel comfortable calling Kepler's laws "laws" is that they can be shown to be part of a bigger system that does not make reference to particular objects or instances. Newton's laws which show why the planets behave as they do are perfectly general. The same is true in evolutionary biology. When we are talking about human evolution out of Africa or whatever we are not applying laws that apply to, and only to, humans. We are using general evolutionary principles to explain what is going on.

Nor, to take one last critical comment, should you be thrown by claims that human evolution refers to unique events – we only got up on our hind legs the once – and hence cannot be subject to laws, which by their very nature demand repeatability. You get unique events in physics, too, and you can explain these. There was one and only one big bang as far as our universe is concerned, and yet people like the physicist Steven Weinberg (1977) can go a very long way to explaining how and why it all happened. The same is true in human evolutionary biology. Scientists must and can abstract from the particular and see it as part of the general if they are to do their work. Suppose humans did only get on their hind legs the once; if you then go on to offer explanations about the ability to cover large distances at a steady walk or jog, or the benefits of not getting overheated because too much sun is hitting your body, obviously you are treating hominins as entities that are subject to the usual laws of kinematics and heat exchange and so forth. You are treating us – and properly so – as things that have features in common with other things, and hence open to explanation by law.

Darwinian theory as science (theory)

We are almost ready to move on, confident that we have made some response to the extreme charges about Darwinism not being a theory. A theory in the sense of a body of laws, that is. Admittedly, although we have addressed the question of law, we have said little about what it is to be a "body." Actually, we touched on this in the first chapter. The traditional answer is that a theory is an axiom system, a hypothetico-deductive system, where laws at the top act as premises to explain (through deductive inference) the laws at the bottom (Braithwaite 1953; Hempel 1966; Nagel 1961). Yet, as many have pointed out, by and large in real science not only is there not even an appeal to major all-embracing laws, there is no attempt to bind everything together in formal systems (Giere 1988). Rather, researchers run up little models of how they think the world runs in particular circumstances – in other words, the exceptions are already built in or avoided – and then they see if they can give the models empirical content. So what we get are clusters of models, joined by the shared focus on certain areas of inquiry and united in the use of the same or similar tools of understanding. This is certainly the case in evolutionary biology (Beatty 1981; Thompson 1989). None of this is to deny that there is what you might call a literary quality, a narrative flavor, to much evolutionary writing. When building models perhaps you would ideally like in these cases, at least, to have formal reasoning, from premises deductively to conclusions. And often indeed that is precisely what you do have. However, the nature of the beast means that evolutionists often have to fill in gaps using their imaginations. Evolutionists offer stories to make the whole picture and to make it convincing. Certainly this is still very much the case for humans. Whether they do this too much is a matter we shall consider shortly.

What about the charge that Darwinism is a theory in the sense of "iffy hypothesis"? This of course is a favorite of the Creationists, and so we really don't need to spend too much time on the worries of people who almost by definition have deliberately removed themselves from the domain of genuine scientific thinking (Ruse 1988a). In the case of evolutionary theory as a whole, we have seen that there are two main sources of evidence. First there is the direct evidence which today includes artificial selection, as in the increase of oil content in corn; experimental evidence, as in the acquiring of alcohol tolerance by fruitflies as shown by Francisco Ayala and his students; natural changes, as recorded by Peter and Rosemary Grant on the Galapagos, as the finches vary and change especially in beak size under the sway of climatic conditions; and human-caused natural selection, as when various pests develop immunity to the killing agents employed by agriculturalists, medical people, and so forth. Second there is the indirect evidence coming from the consilience, raging across sociobiology, paleontology, biogeography, systematics, anatomy and morphology, embryology, and so forth. Critics object that the direct evidence only supports what is often known as "microevolution" whereas what we really need to know about is "macroevolution." We need to know about large-scale changes over large time scales (Gould 2002). I am not sure that this is quite all there is to be said on the subject. Obviously you need large time scales to see if large patterns emerge, but nothing in Darwinian theory suggests that what happens on the large scale is not made up of small-scale causes. *Fallacy*

But what about the large scale? Here we turn to the indirect evidence – evidence that points to the operation of natural selection leading to evolution, even though we do not see it directly. Let us agree that if a consilience of inductions works anywhere, it works here. We have a large range of phenomena from different fields all brought together beneath one overarching hypothesis. Moreover this hypothesis has certainly led to discoveries that were hardly anticipated when first formulated. Think of the molecular homologies between humans and fruit flies, or if you want something more directly selection-related, there is some fascinating work based on the ideas of William D. Hamilton (1967) explaining odd sex ratios. For instance, if brothers are competing for mates, then it is in the interests of a mother to have fewer sons than daughters. From her perspective, one son is like another and it is silly to waste resources on a gang of them. "Local mate competition" has been found to lower male sex ratios in many cases, notably in fig wasps in Central America, where brothers do indeed compete for mates (Herre *et al.* 2001; Ruse 2003).

However, there are those (Popper would have been one) who worry about the whole form of argumentation using the consilience, or its related forms. (The American pragmatist Charles Sanders Peirce (1997) spoke of "abduction." Today, a term popularized by the late Peter Lipton (1991) is "inference to the best explanation.") After all, critics point out, a consilience is not deduction, where the truth of the premises guarantees the truth of the conclusion. Or rather, inferring the truth of the premise in a consilience is not a deduction. It could be wrong. Which is true, but whether in the evolutionary case that matters a great deal is another matter. Logically, the mid-nineteenth-century naturalist Philip Gosse (1857) might have been right. God might have put all of the fossils in place miraculously to give us the impression of evolution. But it is not very likely. The simple fact is that in real life, and in science, we don't demand the absolutely logically necessary, but the highly reasonable. We use consiliences all of the time in court rooms. The butler is convicted of the murder on the basis of the bloodstains and the opportunity and the motive and method and so forth. What we are looking for is something that is "beyond reasonable doubt." And note that often we prefer indirect evidence to direct evidence. In cases of assault, where things happen quickly and there is much stress, eyewitnesses are notoriously unreliable. All of this applies directly to science in general and to biology in particular. The evidence for evolution as a fact – the evidence at least for those prepared to take argumentation and evidence seriously – is beyond reasonable doubt. And the fact that we can think up daft, science-fiction scenarios that say otherwise does not alter the strength of this conclusion. *Archaeopteryx*, the Galapagos finches, the vertebrate limb homologies – the case is beyond reasonable doubt.

Human evolution as science

What about humans? Well, we are not going to run experiments on humans to see if selection has major effects, obviously. There are both practical and ethical objections to doing this. But there is lots of evidence that selection has worked and does work directly on humans. Frankly, in a world where so many go to bed hungry and are ravaged by diseases like AIDS, you would have to be pretty callous to expect otherwise. There are records suggesting that diseases have been important even in Western societies. Thanks to drugs, health policies, and other support, TB is no longer the scourge that it was once. But even in the nineteenth century there was evidence that as people moved to cities the incidence of TB rose, but then started to decline naturally as the tubercular types died early before reproducing (Dobzhansky 1962). A recently reported instance of natural selection in action is yielded by the so-called Framingham Heart Study (Byars *et al.* 2010). This is a massive study of the health features of 14,000 residents of Framingham, Mass., started in 1948. Researchers tracked the medical records of over 2,000 postmenopausal women, checking height, weight, blood pressure. and cholesterol levels against the number of offspring that they had. They found that being a bit on the plump side (skinny women often have fertility problems), with blood pressure and cholesterol not too low, were significant factors in fertility. More than this, they found that there is a genetic movement in the direction of increased weight and shorter height. By 2049 (ten generations) the forecast is that the women will be 2 cm shorter, 1 kg heavier, have healthier hearts, have the first child five months earlier, and reach menopause ten months later than women today. Not that rapid a set of changes, and obviously it could be derailed by any number of factors, but well within the range of many animals and plants.

Microevolution admittedly, but one example of a very rapidly growing number of like examples, now being detected thanks to the success of the Human Genome Project telling us so much about the nature of human genetics: selection for living in high altitudes, selection for staying warm in cold conditions, selection for digesting cereals, and more (Gibbons 2010). So what about macroevolution? We have already seen lots of evidence that if the Darwinian consilience applies to any animal, it applies to the human animal. The fossil record for hominids is now really very good, and grows almost daily. Combine this with the genetic underpinnings and we know a lot about our phylogeny. Likewise with biogeography. (In a later chapter, when turning to race, there will be more to say on this topic.) And we have seen the dazzling results of molecular homological investigations, tying us in with the whole of the animal kingdom. Whether you much like being related to a fruit fly, the fact of the matter is that you are. The evolutionary underpinnings of human social behavior – for a while known as "human sociobiology," although today often hidden under other names – is an area of much controversy. When the field was being opened up in the 1970s, there was much criticism. Some was epistemological, but other charges were more to do with values, with human sociobiology being accused of the sins of racism and sexism among others. I will leave these latter issues until a later chapter, but for the moment I simply acknowledge criticism has not stopped research in the field, or successes being claimed.

One well-known recent piece of work is due to the Harvard-trained biological anthropologist Sarah Hrdy (1999). She has long been interested in infanticide, where for various reasons adults kill infants. There is a lot of (non-human) biological literature on this topic. For instance, it is well documented that in some mammalian species – lions and lemmings, to take extremes – a male taking over a group of females will at once kill all of the very young. This brings the females more quickly back into heat and the new male can fertilize with his own sperm rather than that of others. Not very nice, perhaps, but a very sound strategy from a Darwinian perspective. We get infanticide in humans, too. For instance, it is far from unknown for very poor and socially ostracized mothers to kill their infants. This is the stuff of novels, like Hetty in *Adam Bede*, and again can be a good strategy if it makes possible shortly a more successful mating where some help or status might be offered. Hrdy focuses on the fact that some groups or castes in India – those with high status – have an offspring sex ratio that is very heavily weighted toward males. Even if one discounts today's possibilities of pre-birth screening and selective abortion, the fact is that girls in these groups often do not survive – sometimes, virtually none survive. Why is this? Why are parents, directly or indirectly through neglect, killing their daughters? Hrdy draws attention to a well-known law or rule, the Trivers–Willard rule, that has been found to hold in many species of animals (Trivers and Willard 1973). High-status mothers tend to have sons. Low-status mothers tend to have daughters. It is found again and again, for instance in the large South American rodent introduced into Britain, the coypu. Mums up the scale have boys. Down the scale, girls. This is not a one-off, but grounded firmly in Darwinian theory. Males can have lots of offspring – they produce a lot of sperm and, especially if like most animals have little or nothing to do with parental care, can in theory have many offspring. Females are left holding the baby, literally, and their offspring production tends to be limited, even for the most successful. However, the flip side is that some males will have no offspring, whereas pretty much every female is guaranteed some offspring. Hence a female who can give her sons a good start in life is probably better off investing in sons whereas a female who cannot do this is better off investing in females, because she knows that at least some (probably all) will reproduce – not necessarily well, perhaps, but better than sons with nothing. This is anthropomorphic language, and no one suggests that something like the coypu is doing any serious planning. But Hrdy argues that the human case is analogous, in that higher-status people are better off with sons and lower-status with daughters, and this is what happens in the Indian case. The wives of the high-status sons come from lower down the social scale, but are then raised up by their husbands and their families. Significantly, the low-caste families invest heavily in dowries for the daughters, but tend to let the sons fend for themselves.

Final causes and adaptations

This example shows an interesting facet of evolutionary explanation. We have two levels of explanation here (Mayr 1988; Ruse 2003). On the one hand we have *proximate* causes at work. Presumably for the coypu we have hormones and such things triggering the effects. For the humans, presumably we have social customs and so forth. Then on the other hand we have *ultimate* causes: why does this happen? In both cases, to maximize reproductive chances by adjusting the benefits given to children of different sexes. Ultimate causes speak to ends. They tell us what the purpose or function of something is, and as we saw in chapter 1 they go back to Aristotle (if not to Plato), who labeled them "final causes." A more recent term is "teleological explanation." In this respect, as was shown by our coverage of the history, Darwinian evolutionary biology (not just human Darwinian evolutionary biology) is very different from physics and chemistry, where final causes were expelled 400 years ago during the Scientific Revolution. No one asks what the purpose of the tides is, for instance, although we might well ask about the proximate causes of the tides. You can certainly ask: what is the purpose of the hand or the eye?

Although, as we saw, Darwin strove mightily to give a naturalistic explanation of such final-cause-demanding phenomena, things which he often called "contrivances" and that are today generally called "adaptations," there are some philosophers (and biologists) who are made very uncomfortable by the presence of final causes and argue for their elimination (or at least that we conceal them under such names as "teleonomy"). They say: "If Darwin showed how eventually everything can be explained using proximate causes, then let us stick to this. Let us eliminate talk of purposes and functions completely from our language." But is such linguistic genocide either desirable or possible? The nub of the problem here – the focus of the dispute (for not all think we have a problem) – is that we are working with a metaphor. We are looking at the organic world as if it had been designed, and since in human design it is appropriate to ask about ends or purposes – what is this strange-looking instrument that you have in the kitchen drawer? What do you use it for? – we think it appropriate to ask about ends or purposes in the biological world. But other than for religious believers – and even for believers when they are doing science – we don't want to bring intentional purposes into the organic world. That is the whole point of Darwinian selection, to avoid having to do this. So what we have are final causes without a designer, in other words we are thinking metaphorically.

Some philosophers just don't like metaphors. They look upon them as a sign of weakness or immaturity. The seventeenth-century English philosopher Thomas Hobbes was one. The twenty-first-century American philosopher Jerry Fodor is another (1996, 2007). He says that as science matures, you throw out the metaphors and bring in the mathematics. This seems to me (and I should say a great many other people, especially those who take literary theory seriously) just plain silly (Ruse 2005a). I don't see how you can do science without metaphor – work, pressure, attraction, genetic code, struggle for existence, natural selection, Oedipus complex. At the very least, metaphors have incredible heuristic value. No one could have discovered – no one could have asked – about the purpose of the plates on the *Stegosaurus* without the metaphor of design. They certainly could not have come up with the answer that they are for heat control. Whether you like Hrdy's explanation of Indian infanticide or not, it is a phenomenon in need of explanation and she does offer one.

A more specific worry concerns this particular metaphor of organic design. The classic problem with teleological explanation is that of the missing goal object (Mackie 1966). You explain the eye in terms of the end of seeing. But what happens if you get something in the eye early on and are blind for life? The eye hardly exists for seeing! Note that this is not a problem for proximate causes, because they always come before or at the same time as the effects – I slam the door on my thumb and start to swear. If you are dealing with conscious design, you get around the problem by talking in terms of intentions. I made the kitchen instrument to take the stones out of cherries and, even though my wife turns out to be allergic to cherries and we never eat them, the intention (which is a proximate cause) still holds. In the case of biology, the best thing is to point out (as did the great German philosopher Immanuel Kant [1790] (1928)) that in the case of biological function, we have a kind of chain situation. Situation of type A causes situation of type B, and then B causes an A. Eyes lead to seeing, and then the seeing leads (via reproduction) to more eyes. We just assume that things will generally work out fine, and so (based on the past) we can make a prediction about what will happen. Sometimes it just doesn't work out and that is too bad, we were wrong. But there is no objectionable messing around with causes and effects and time, like supposing that causes in the future are affecting what is happening in the present. It is rather that we refer to the future (or what we think will be the future) in trying to understand the present.

Agree that this holds just as well for human explanations as for the rest of biology. There are still critics. Remember that the late Stephen Jay Gould accused Darwinians of spinning "just so" stories, akin to the tall tales that the British novelist Rudyard Kipling used to tell, to amuse children, such as that the elephant got a long nose because a crocodile seized it one day when the elephant was drinking and the nose grew longer and longer as the elephant tried to escape. Of course these are just silly, made-up fantasies. Gould thought that Darwinians dealing with humans offered particularly egregious examples of this practice. Gould's charge, one that a lot of philosophers think well taken, is that Darwinians are no better than Kipling when they are trying to come up with adaptive scenarios, and that in the case of humans Darwinian creativity goes into overdrive (Gould and Lewontin 1979).

There was no one like Gould for seizing on a metaphor or analogy and pushing it to the extreme, making his opponents seem flatfooted and stupid. This is a clear case in point. No one could deny that Darwinians have often gone off on unrestricted flights of fancy, seeing adaptation

everywhere and spinning tales to explain it. But note that there is a difference between simply trying to be the biological equivalent of Rudyard
Kipling or Hans Anderson and using one's imagination in a way that is
absolutely vital to creative science. As Popper (1963) rightly used to stress, a
scientist has to come up with hypothesis after hypothesis, trying to understand nature. But then these must be put to the test – that is the difference
between inventing fables and doing science. Conjectures and refutations,
as Popper used to say. Let a thousand flowers bloom, as Mao used to say. (He
was pretty good at cutting them down!)

Go back again to our *Stegosaurus* case (Farlow *et al.* 1976). Paleontologists
came up with hypothesis after hypothesis to explain those silly-looking
plates, including one which no longer had the plates upright but lying flat
and pointing out away from the body. The most popular early hypothesis
was that they were needed for defense – the herbivorous dino protect-
ing itself against all of those meat-eaters. But this obvious suggestion ran
into trouble when people started to look hard at the plates themselves
and the points of attachment to the body. They just didn't seem strong
enough. They would crack and break at the first attack. So then the suggestion was made that the plates were produced by sexual selection, less
like the antlers on a deer (which are used for fighting) and more like the
feathers on a peacock, which are used for attraction. But the trouble here
is that all of the stegosauruses seem to have the plates, so you do not get
the male–female asymmetry (sexual dimorphism) that you expect from
sexual selection. Then came the heating and cooling hypothesis, clearly
inspired by the fact that the plates of the stego look like the plates or fins
that you find in electrical generating plants (especially in the cooling towers). But evolutionists did not stop here. They went out and looked for confirming evidence, which they found in such things as the way the blood
would have flowed from body to plate and back again – as well as other
points of support, for instance working out the amount of heat produced
(that would need to be dissipated) by the consumption of vegetable matter
and the environments in which the brutes lived and from there working
out the expected air flow and how the plates could handle it adequately.

What about the human case? Take a nice example from long back in our
history (more than 50 million years ago), when the primates first appeared.
These animals had two distinctive characteristics, grasping hands and feet
and stereoscopic vision (the ability to see in 3D). Several models have been

offered to explain these features. One is that it is all a question of being up in the trees (Szalay and Dagosto 1988). You need the hands and feet to grasp the branches and the 3D vision to judge distances. Another suggestion is that hands, feet, and eyes evolved in tandem to catch insects (Cartmill 1974). Suggestively, other animals like hawks and cats have stereoscopic vision. These are all active hunters. A third suggestion is that the hands came first, as the proto-primates gathered food (fruits and nuts) rather in the fashion of squirrels (Susmann 1991). Then the move to the other features came as the primates turned to trees as homes and safe havens. All of these start off as "just so" stories in a Gouldian sense, but they did not stay that way. They have certain (different) consequences that can be tested. In particular, the first two hypotheses see the features all coming together, whereas the last hypothesis sees hands first and 3D vision down the road. The discovery of a skeleton of *Carpolestes simpsoni*, a primate-like creature from about 55 million years ago (in what is now Wyoming) throws light on these claims (Bloch and Boyer 2002). Pertinently, it has proto-primate claws and feet (including an opposable big toe), but it has eyes on the side of the head rather than looking forward in tandem. In other words it did not have stereoscopic vision. So the first two hypotheses must be wrong, and at the moment the laurels go to the fruit-gathering-and-only-later-3D-vision hypothesis. Now the problem is to track down in more detail how that happened.

Note that it is no criticism of the science that evolutionists do not yet have the full answer. There are lots of places where we are still in doubt. Remember the discussion of bipedalism. There is still considerable doubt about why and how this happened. But it is not a cause for despair, although Creationists and fellow-travelers delight in pointing to problems and unanswered questions. It is in the nature of science that you start the day with a problem, solve it by lunchtime, and then have two more problems by the time you pack up for the day. If it weren't, there would be no point or thrill to science. No one wants to spend their days just polishing or cleaning finished theories. In the bipedalism case, for instance, the newly discovered and reconstructed *Ardepithecus* is already playing a major role, as paleoanthropologists try to work out its full significance – the nature of the bones themselves, the climate and habitat in which it lived, its relationship to other hominins like the australopithecines, and so forth. We know that some things don't work, like the went-out-on-the-prairie-and-got-up-on-its-hind-legs hypothesis, but we certainly don't yet know fully what does work.

Spandrels

Gould had not yet finished with Darwinism and its extension to humankind. Related to the "just so story" critique was another one about Darwinians being overly fond of adaptive hypotheses, seeing natural selection as working flat out and successful everywhere. He thought that we were living with an outmoded natural theology, and that we should not expect to find efficient functioning or final causes throughout the living world. Remember (from chapter 1) how he accused Darwinians of being too optimistic about the workings of nature, and of being Panglossian, a term he took from the name of the philosopher in Voltaire's *Candide*, a man who (following the philosopher Leibniz) saw this as the best of all possible worlds. To the contrary, borrowing a term from church architecture, Gould argued that much of the living world is made up of spandrel-like features, where he was referring to the non-functional triangular areas at the tops of pillars in medieval churches – often used for decoration but without any structural purpose. Backing this claim, Gould mentioned not only the usual suspects (causing non-adaptive features), like genetic drift, but also something of which he made much: "constraints." Influenced by German thinking, Gould argued that in building organisms there are all sorts of factors that need to be taken into account, and satisfying these factors cut down on the opportunities for full adaptive "optimization." As in real life, you have to be satisfied with what you can get, not what you would want. Thus for instance you cannot make a cat as big as an elephant, because as height is linear, weight is cubed, and a cat of elephantine proportions would never have the delicate legs that it needs for its jumping and running predatory lifestyle. In fact, Gould pushed the argument even further, arguing that constraints from the past often get embedded in the ground plan (or *Bauplan*) of the organism, and today we live with long-forgotten and quite unnecessary basic features. He instanced the four-limbedness of vertebrates as a case of a constraint from the past, today quite without adaptive significance.

A huge amount of ink has been spilled in the past thirty or forty years, since Gould first started to make these kinds of charges. There is no need to go over them in detail here. (See Ruse 2003 for a full discussion.) On the one hand, Darwinians point out that from the *Origin* on they have always agreed that natural selection cannot do everything and that not every feature of the organic world is going to be perfectly designed. On the other hand, a lot of Gould's charges are being blown up way beyond what is justifiable, and when looked at in detail start to seem less threatening than appeared at first. Take the matter of constraints and agree that these are important. There are reasons to do with physics as to why an elephant has elephant-legs and not cat-legs (Vogel 1988). But note that things are often not quite this clear-cut. The four-limbedness example is a good case in point. For a start, when four-limbedness first appeared, vertebrates were aquatic and a major need was to move up and down in the water. The late British evolutionist John Maynard Smith (1981), who in earlier life had been an aero-engineer, pointed out that their problem was similar to that of planes needing to move up and down in the air, and the way in which they do this is by having two wings at the front and two at the back. In other words, four-limbedness was adaptive when it first appeared. It was not a constraint for there were other vertebrates that experimented with other combinations – just as we have now discovered organisms with eight or nine digits, so it is a little misleading to say that having five digits (another popular example) is a constraint. So it is all a little misleading simply to say that something like four-limbedness is non-adaptive now and leave it at that. Moreover, when you think about it, it is not so obvious that four-limbedness is a constraint even now. Snakes have no limbs, ostriches have two, humans four, and elephants five. Why exclude the nose as a potential limb?

Move on now to consider the specific human case. In one sense, it would be ludicrous to deny the effects of natural selection or to forswear talk about adaptation. Hands, eyes, teeth, and so on and so forth are clearly adaptations. That is beyond doubt. Having said this, we are not perfectly adapted and there are all sorts of cases where rival forces might have been at work, where there are clear constraints, and where there are compromises, with both sides having to give a little. Take the facial hair of human males. Is this simply a function of the testosterone levels that males need in order to function as males, or is it in some sense pulled along by sexual selection? I am not sure that anyone has an answer to this, any more than they have an answer to the biological function (or not) of male baldness.

Take again human height. Being taller is not necessarily adaptive, but one can well see reasons why height does give its advantages. There are obviously constraints here. The taller you get, the more physics starts to kick in. It is well known that people over seven feet tall tend to have circulatory problems, a simple function of pumping the blood up to ever-greater heights. And human childbirth is a classic case of the need to compromise biologically. The bigger the head at childbirth, the quicker you can get moving on to providing all of that adult brain power. But the bigger the head at childbirth, the more difficult it is to expel the baby from the mother. Clearly natural selection has had to move toward what is not always a very easy balance between two demands. (There is more on these sorts of things in chapter 8.)

Culture

What specific issues come up as we move into the more distinctive human realm, and in particular what comes up as we move toward culture? Let us work against the background of two basic assumptions. No sensible person is going to deny that human culture has an adaptive value. We have exploded in numbers on this Earth of ours because of our culture, and we survive and reproduce because of our culture. I lived for forty years in Canada. Without culture, no human being is going to live in Canada. Gould (2002) did argue that culture is all what he called an "exaptation," something which goes off on its own course (perhaps with its own function) without being connected to its origin or first function. But truly only an extremist would deny that culture is important to human biological flourishing. The second assumption is that it would be silly to tie culture too closely into adaptation. The whole point about culture is that it is so very flexible and that it can run off in ways that surely do not help survival and reproduction. One thinks of eating habits in North America today. No doubt in the past it was a good thing to stock up on fats and sweets. Without doubt it is not necessarily a good thing today to stock up on fats and sweets. This sort of example can be multiplied without end, perhaps even to the ultimate end. One can certainly see how a grasp of science is culturally advantageous. You can live in Canada through the winter, thanks to modern science and technology. But if at some point in the future we blow ourselves all up with the Bomb, then so much for adaptive advantage.

Controversy starts to arise as one tries to move beyond virtual truisms like these. Researchers are all over the place, and although, as noted earlier, no one seems much inclined now to call themselves a "human sociobiologist" – perhaps understandably since there was so much controversy over sociobiology back in the 1970s – there seems not yet to be any one, fixed, alternative name. Perhaps any evolutionary approach to human nature could properly be called "evolutionary psychology," but as it happens this name has been appropriated by one of the more controversial schools or approaches (leading some commentators to refer to this approach as "Evolutionary Psychology" leaving the uncapitalized form for the generic approach). These Evolutionary Psychologists, prominent among them being the husband and wife team of John Tooby and Leda Cosmides (2005), spell out a number of principles to which they are committed in their research. These include an ardent adaptationism, combined with the cognitive science belief that the mind is a computer of some kind, designed (by natural selection) to extract information from the environment and to enable us humans to manipulate and live in the environment (which is composed, in large part, of our fellow humans); the belief that the programs being run by the brain are themselves adaptive, having been put in place by selection; a commitment to a form of mind/brain modularity, that is a belief that the brain is not just an all-purpose computer but can be broken down into parts or modules, each of which was fashioned by selection and acts in the adaptive interests of its possessor; the belief that much of the human brain was put in place in the Pleistocene (from 1.8 million years ago to 10,000 years ago) when *Homo sapiens* was evolving, and hence might not be directly or even indirectly adaptive today; and generally a commitment to some kind of universality, that is a belief that the genetic differences between humans are not that significant when it comes to the mind and the ways in which it fashions culture.

Obviously not all of these beliefs are unique to the Evolutionary Psychologists. There are many who are committed to adaptationism in one form or another, including adaptationism as applied to the human mind and behavior. In the last chapter, brief mention was made of the modularity thinking of the archeologist Steven Mithen (1996). And to take another point, many linguists would agree that the genetic differences between humans cannot be that crucial. The whole point about language is that it has to be a shared adaptation. There is no point in my speaking perfect English if the rest of you are so far over the place with dialects that not one of you can understand me (or each other). What can be said, however, is that bringing these various factors together in one cluster has made the Evolutionary Psychologists' program deeply suspect in some circles. In the philosophical world, for instance, it is hard to find even a disinterested exposition of the kind I am trying to give here. (See, for example, Buller 2005). Why the violent dislike? Overall, there is the sneaking suspicion that within even the most hardened of secular naturalist breasts there beats a heart that fears that a biological approach to the deepest and most human aspects of our nature conflicts with yearnings that we humans be judged special. There is not a lot one can say about this, other than to note that some of the most violent critics of Darwinism-as-applied-to-humans have subscribed openly to nineteenth-century Germanic philosophies that elevated the status of humankind. (Richard Levins and Richard Lewontin [1985], who make their enthusiasm for Marxism a central part of their thinking about evolution, are good examples.) It may be that the discussion in the next chapter will add to our understanding of this point.

A more specific reason for the opposition to Evolutionary Psychology centers on adaptation. Anyone in the Gould tradition (and his language seduced many on the humanities sides of campuses) who does not like adaptationism is going to be in the band of critics. Analogously, the modularity thesis is sometimes pushed to extremes. For instance, one claim that has been made in the name of Evolutionary Psychology is that men get sexually attracted by women with an ideal waist–hip ratio (this being 0.7), supposedly a good indicator of fertility (Singh 1993). It is hard to imagine that there is a part of the male brain that is specifically and exclusively devoted to this passion. Having said this, of course, one might not want to trash the modularity thesis entirely. Apart from the fairly detailed knowledge that some parts of the brain are used for and only for certain functions – one thinks of the brain parts used in language already mentioned in the last chapter, and later we shall see (in chapter 7) evidence about one specific part of the brain and its causal role in sexual orientation – there is the analogy of the rest of the human body. We are not an all-purpose machine nor are our parts totally interchangeable. We are mid-range primates, much given to sociality, and our hearts pump our blood, our lungs breathe the air, our mouths take in our food, and so forth. Joking aside, by and large we do not use our lower orifices for speaking. Why should the brain be any different?

What about the claim that today human biology (in terms of mind and culture) is not something really flexible and to be considered in its own right, but rather that in trying to understand it we must refer to the past. Must we agree with the Evolutionary Psychologists that we should base our thinking on the Pleistocene era? Must we go along with the argument that, judging from the fossils and from culture and so forth, that was when the human mind in its present form came together and looking for significant changes in the past 10,000 years is really pointless? Evolution is just not that fast. Not everyone is convinced that we must make this move. We have seen that biological evolution is still occurring in humans, and the critics seize on this point. They argue that although 10,000 years is indeed a short time in evolutionary terms, it is not that short. There is good evidence, for example, that with the coming of agriculture, many humans developed lactose tolerance – in fact, just those with a history of cattle farming. Why should this not also be the case for other features, including those more directly affecting culture? (Although of course lactose tolerance does influence the keeping and farming of certain animals rather than others.) And in any case, culture can seriously affect the uses that we make of our biology.

Philosopher Stephen Downes (2010) focuses on the human hand, arguing that its functioning is going to be a major factor in brain evolution. Thus, on the one side, since well before the Pleistocene the hand was used for purposes like grasping and tool manufacture and use, we should expect to find the brain reflecting these ends. On the other side, the hand is now used in ways far more sophisticated than were thought of in the Pleistocene – for making fishing nets, for example. Are we to say that there was no corresponding change in the brain? "Crucially," writes Downes, "the brain's evolution is determined, in part, by the hand. The more fine-grained motor skills our hands are capable of, the more motoring of these skills is required by the brain." He goes on to say that: "The full story of the evolution of the human mind will include components that account for our lineage's divergence from that of the rest of the great apes and components that account for our predictable psychological responses to aspects of large highly complex social groups" (Downes 2010, 250). In other words, we need to take account of our more distant past and our more recent past as well as the Pleistocene.

We are not going to resolve the debate here. In later chapters, we shall again encounter the Evolutionary Psychologists, for they have had things to say (very interesting things to say, whether true or false) about the nature of knowledge and of moral behavior. My sense (against the philosophers) is that they offer us a stimulating and fruitful method of inquiry, and (against the enthusiasts) that Evolutionary Psychology is surely only one part of the picture. Indeed, we have ourselves already encountered a somewhat different approach (Buller 2009). Sarah Hrdy falls more naturally into the group known as the "human behavioral ecologists." They stress the flexibility of adaptations. You are not stuck always doing the same thing in the same way, whether or not it is adaptive. Human nature responds to the situation, and different situations elicit different responses. Thus Hrdy does not assume that in India there are genetic differences between the castes. Rather, she thinks that if you are born upper-caste then you will adopt one strategy (favor the boys) and if you are born lower-caste you will adopt another strategy (favor the girls). This is clearly as adaptationist as Evolutionary Psychology, but without some of the other more provocative claims or restraints on explanation.

Memetics

Biology is the starting point both for Evolutionary Psychology and for human behavioral ecology. What about cutting out the biology completely and considering culture *sui generis*, as it were? One such approach that has garnered much attention recently, particularly in the popular press, is the so-called theory of "memetics." This started with an idea thrown out by Richard Dawkins at the end of his *The Selfish Gene*.

> The gene, the DNA molecule, happens to be the replicating entity that prevails on our planet. There may be others. If there are, provided certain other conditions are met, they will almost inevitable tend to become the basis for an evolutionary process.
> But do we have to go to distant worlds to find other kinds of replicator and other, consequent, kinds of evolution? I think that a new kind of replicator has recently emerged on this very planet. It is staring us in the face. It is still in its infancy, still drifting clumsily about in its primeval soup, but already it is achieving evolutionary change at a rate that leaves the old gene panting far behind.

The new soup is the soup of human culture. We need a name for
the new replicator, a noun that conveys the idea of a unit of cultural
transmission, or a unit of *imitation*. "Mimeme" comes from a suitable
Greek root, but I want a monosyllable that sounds a bit like "gene". I hope
my classicist friends will forgive me if I abbreviate mimeme to *meme*. If it
is any consolation, it could alternatively be thought of as being related to
"memory", or to the French word *même*. It should be pronounced to rhyme
with "cream".

Examples of memes are tunes, ideas, catch-phrases, clothes fashions,
ways of making pots or of building arches. Just as genes propagate
themselves in the gene pool by leaping from body to body via sperms
or eggs, so memes propagate themselves in the meme pool by leaping
from brain to brain via a process which, in the broad sense, can be called
imitation. If a scientist hears, or reads about, a good idea, he passes it
on to his colleagues and students. He mentions it in his articles and his
lectures. If the idea catches on, it can be said to propagate itself, spreading
from brain to brain … When you plant a fertile meme in my mind you
literally parasitize my brain, turning it into a vehicle for the meme's
propagation in just the way that a virus may parasitize the genetic
mechanism of a host cell. And this isn't just a way of talking – the meme
for, say, "belief in life after death" is actually realized physically, millions
of times over, as a structure in the nervous systems of individual men the
world over. (Dawkins 1976, 206–07)

Some have taken up this idea with enthusiasm. The philosopher Daniel
Dennett is one. At a general level, he thinks memetics is a good tool for
understanding the spread of ideas. He stresses how fast ideas can move
and how difficult it can be to stop them. "Memes now spread around the
world at the speed of light, and replicate at rates that make even fruit flies
and yeast cells look glacial in comparison. They leap promiscuously from
vehicle to vehicle, and from medium to medium, and are proving to be
virtually unquarantinable" (Dennett 1990, 131). We must remember that
memes serve themselves, not us. Some memes nevertheless are good memes
from our viewpoint. These include general memes like cooperation, writing, and education, and more particular memes like *The Marriage of Figaro*,
long weekends, and undergraduate majors. Other memes are less desirable
from our perspective but perhaps very successful from their perspective.
Included here are anti-Semitism, computer viruses, and spray-can graffiti.
At a more specific level, namely the arts and music in particular, Dennett

argues that memetics throws significant light on things like tunes stick-
ing in our minds even though often we wish we could drop them. He notes
that there are various (what we might call) adjunct phenomena around
memes, for instance positive feedback. In the biological world, this can
lead to such phenomena as "runaway sexual selection," when – the peacock's tail being a possible example – something feeds on its own success
and gets bigger and bigger. Following Dawkins, Dennett mentions best-
sellers (something both he and Dawkins have good experience of!), where
once a book starts selling and getting on bestseller lists, it too feeds on its
own success and sells more and more copies.

Turning to criticism, both philosophers and scientists have been less
than overwhelmed by memetics. Cambridge philosopher Tim Lewens
(2007) has been particularly critical. He does not think that memes are
proper replicators in the ways that genes are in the biological world. Sure,
often we copy an idea, but how faithfully it gets copied is another matter.
A recipe for a cake, for instance, is almost invariably modified from one
kitchen to another. In any case, an idea is often not copied directly, but
the spread occurs because people go back to a common source. To quote
the cognitive scientist Dan Sperber, "most cultural items are 're-produced'
in the sense that they are produced again and again – with, of course, a
causal link between all these productions – but are not reproduced in the
sense of being copied from one another ... Hence they are not memes,
even when they are close 'copies' of one another (in a loose sense of 'copy',
of course)" (Sperber 2000, 164–65). Lewens does not think that we get the
lineages that we get in genetics and that make it possible to formulate
laws of transmission like those of Mendel. Instead of an idea having just
one parent or source, we often get the idea from multiple sources, and it
is only as they come in that the idea lodges firmly in our minds. Perhaps
in some cases one person uniquely has a notion which is then passed on,
but more often in culture it is a group experience. Finally, Lewens questions whether the very notion of a cultural unit like a meme makes much
sense. In the words of anthropologist Adam Kuper: "Unlike genes, cultural
traits are not particulate. An idea about God cannot be separated from
other ideas with which it is indissolubly linked in a particular religion"
(Kuper 2000, 180). To take an example of Dennett's, is an undergraduate
major in philosophy one meme and an undergraduate major in physics
another, or are they both manifestations of the same meme? Perhaps in a

case like this one is splitting hairs, but in dealing with a major issue like anti-Semitism, it is surely going to be important to ask questions about whether Luther's anti-Semitism was the same as Hitler's and if it is not whether there is nevertheless some connection. Without proper clarification on matters like this, discussion is unlikely to advance very far.

Computer scientist Bruce Edmonds (2002) is another who is very critical of the meme notion. His objections, as a working scientist, are less conceptual than pragmatic. Where is the payoff for adopting memetics? He complains that the notion is all talk and very little action. He fears that we are simply redescribing things in the fancy language of memetics. If the theory is to have any real explanatory value, then it needs to start generating predictions that can be tested and evaluated.

> In my opinion, memetics has reached a crunch point. If, in the near future, it does not demonstrate that it can be more than merely a conceptual framework, it will be selected out. While it is true that many successful paradigms started out as such a framework and later moved on to become pivotal theories, it is also true that many more have simply faded away. A framework for thinking about phenomena can be useful if it delivers new insights but, ultimately, if there are no usable results academics will look elsewhere.
>
> Such frameworks have considerable power over those that hold them for these people will see the world through these "theoretical spectacles" (Kuhn 1969) – to the converted the framework appears necessary. The converted are ambitious to demonstrate the universality of their way of seeing things; more mundane but demonstrable examples seem to them as simply obvious. However such frameworks will not continue to persuade new academics if it does not provide them with any substantial explanatory or predictive "leverage". Memetics is no exception to this pattern.
>
> For this reason I am challenging the memetic community of academics to achieve the following three tasks of different types:
>
> - a conclusive case-study;
> - a theory for when memetic models are appropriate;
> - and a simulation of the emergence of a memetic process.

To date, complains Edmonds, not one of these three tasks has been tackled in a serious way, let alone passed conclusively. Hence, although this is hardly the end blow for memetics, until it starts seriously to produce results we can put it to one side.

Cultural evolution

How about an approach that tries to steer a middle path? One takes biology as basic – Darwinian selection theory is the overall paradigm. But then one tries to tie culture into this, as an adaptation in some sense, but clearly an adaptation of a very innovative (not to say strange) kind – indeed, as noted earlier, an adaptation that might well backfire and from a purely biological viewpoint prove maladaptive. A number of people have tried to work in this mode, offering models and theories to explain how culture operates within a Darwinian framework, how it can be a very powerful form of adaptation and yet how sometimes from a biological perspective it can backfire. Peter Richerson and Robert Boyd (2005), two major players in this cultural evolutionary game, using a simple model of Midwestern farming communities, offer a basic picture of how culture might operate and interact with biology. They invite us to consider two cultural variants of Midwestern farmers. (This is based on real-life cases, but cleaned up and simplified for the purposes of exposition.) On the one hand we have the "Yankee" farmers. They are descended ultimately from British and related stock, they are hardworking and value their lifestyle, but are open to innovation and to commercial possibilities. The land as such is a tool for their ends, they will rent, and if the offer is right, they may well sell. On the other hand, we have the "German" farmers, who value family and land and continuity. The all-important thing is to keep the farm in the family, and outside innovations and ideas come a bad second. Expectedly, the Yankees make more money and work less, but their lives are more fluid. The Germans make less money and work more, but the satisfactions lie in the sense of family and community.

For what one suspects are fairly obvious reasons, the Germans tend to have somewhat larger families – based on real information, Richerson and Boyd give (average) figures of 3.3 children per family for Germans and 2.6 children per family for Yankees. In other words, all other things being equal, the Germans are biologically fitter than the Yankees, and their numbers from generation to generation will show this. But now factor in culture in another way. Kids pick up ideas – as often as not from their peers as from their parents. Again, for fairly obvious reasons, one can expect more being influenced by (seduced over to) the Yankee lifestyle than to the German lifestyle. In other words there will be a transmission of cultural values and ideas – Richerson and Boyd stress that they are flexible as to exactly what this means and do not demand or expect a simple unit notion, as with memetics – and this could be biased in one direction rather than another. So what happens? Obviously, it all depends. It could be (what seems in fact to be true) that the bias in favor of Yankee lifestyle is not so strong as to stop the increase in German lifestyle. It could be that the bias is so strong that Yankee lifestyle starts to wipe out German lifestyle even though German lifestyle has the purely biological edge. Or one could get some kind of equilibrium, with the culture balancing out the biology.

This all makes it sound as though the culture is essentially opposed to the biology, and Richerson and Boyd are at great pains to say that this is not so at all. They argue that culture depends above all on imitation. Of course innovation is very important, but the strength of culture is that a good idea can be shared and spread. Change can occur quickly. This ability to imitate is something that is essentially human, not really found that much in other organisms, including our ape cousins. It requires a huge amount of brain power, which of course we have and others don't. But as the authors stress, brain power is extremely expensive – you need lots of protein (humans use 17 percent of their energy on their brains as opposed to 3 percent for other mammals) – and there must therefore be (or have been) massive selective pressures in its favor. Richerson and Boyd point out that just when human brain power was exploding upward, the Earth's climate was very uncertain, and they argue that a clever ape with a powerful new method of transmitting information would be at a great adaptive advantage. Responses to the changes in the environment could be efficient and very rapid. The costs of extra brain power are met by the virtues of culturally based, shared innovations to tackle nature's challenges.

Of course, biology could not just leave it to chance about how good new cultural innovations might be spread. The suggestion is made that there are at least two significant biologically based aspects to human nature that make for the significant spread of worthwhile new ideas or techniques. First, there is pressure to imitate the common type.

> Recall the old saw "When in Rome, do as the Romans do." This strategy makes good evolutionary sense under a broad range of conditions. A number of processes, including guided variation, content bias, and

natural selection, all tend to cause the adaptive behavior to become more common than maladaptive behavior. Thus, all other things being equal, imitating the most common behavior in the population is better than imitating at random. (Richerson and Boyd 2005, 120)

What is being said here is that, by and large, what is done by others is worth following. Don't deviate for the sake of it. If it ain't broke, don't fix it.

Second, there is the value of following those who are successful. Clearly the successful are using ideas or techniques that pay off. So do likewise. At an almost trivial level, we see this strategy in the ways in which people try to imitate the celebrities in our society today. But, silly as this often is, there is a valuable message beneath.

> Mass-media celebrities notwithstanding, our attraction to the successful makes much adaptive sense. Determining *who* is a success is much easier than to determine *how* to be a success. By imitating the successful, you have a chance of acquiring the behaviors that cause success, even if you do not know anything about which characteristics of the successful are responsible for their success. (Richerson and Boyd 2005, 124)

Finally, one can readily see how maladaptive ideas can as it were hitchhike on the back of a gene–culture process like this. All you need is for some prominent person to have a stupid idea that is nevertheless very appealing – often such ideas do appeal precisely because they are stupid, simplistic, and shallow – and such an idea or practice can spread like wildfire. The fear of witchcraft is a case in point, for it obviously appeals to people who for various reasons feel frightened or threatened, it is something often reinforced and promoted by the people in positions of authority, and yet it can have a corrosive and deadly effect on society, wiping out a whole segment before it runs its course. Something like this shows that, as always in Darwinian biology, it is a matter of compromises. Culture can have deadly, that is highly maladaptive, effects. The point is that overall the positive adaptive values of culture outweigh these bad effects. It is just the same as the compromise between walking upright – a good thing – and having pain-prone backs – a bad thing.

To say that the outlines of this approach to culture are fairly obvious is not to belittle the care and attention that Richerson and Boyd put into developing their insights. There is an attractive, modest sobriety that is often missing from the grandiose claims made by partisans of other approaches.

The big question obviously is that of truth, that is to say the evidence in its favor. The authors themselves admit to the difficulty of finding anything very definitive, especially since much of their speculation rests on events long past. This is not to say that they are simply building "just so" stories in a Gouldian sense. They draw on psychological tests and experiments to support their beliefs about human inclinations to conformity and to respect for authority figures or successful societal members. They look at accounts of different cultures in the past and in varied environmental conditions. They and others are starting to build models to make testable predictions about how culture can be transmitted and developed. Perhaps not surprisingly they seem to have something of a love–hate relationship with approaches like Evolutionary Psychology, sometimes regarding the work of others as something to be appreciated and used and sometimes to be rejected or ignored. Interesting and stimulating but thus far unproven is probably the best judgment.

Religion

Such an assessment leads naturally to the closing discussion of this chapter. Even a sympathetic reader will realize by now that not only is there no one standard approach to the problem of evolution and culture, but that the approaches that have been proposed are diverse and often, if not exactly convincing, then at most not terribly well developed and supported. There are obvious reasons for this, and we have touched on some of the most prominent. Apart from the theoretical, practical, and moral difficulties of dealing with humans – we are after all pretty complex organisms – in dealing with culture we are dealing with something that is in key respects so ephemeral. How can we really make claims about culture (say) 100,000 years ago, or even 50,000 or 25,000 years ago? We may have bones and so forth, but then we are left trying to make sense of a few tools and cave paintings. Scientists are clever mammals, experts at extracting information where there seems to be but detritus and decay, and who dares say that no new techniques will ever come on board? Think of the transformation brought about by our understanding of DNA. But there are limits to everything.

This feeling of near-bewilderment, certainly of frustration, only increases the more one digs into the literature. For instance, there has been much recent interest in religion. After all, today, there is no more pressing cultural issue. Many, probably most, people in the world are religious in one way or another. They believe in a God and an afterlife and a specified morality – or if not all of these, then some and perhaps other commitments and behaviors. If questioned, most would surely say that not only do they think religion to be true, but they think it one of the greatest and most valuable aspects of human culture. Without religion, life would be worth far less. On the other hand, there is a group of people, very vocal, arguing that religion is one of the worse things ever to afflict humankind. Led by Richard Dawkins, the British biologist and popular science writer (remember, the *Selfish Gene*), these "new atheists" argue that religion is something that is evil in every respect and should be eliminated. (See Dawkins 2007; Dennett 2006; Hitchens 2007.)

We know well that historians and sociologists, as well as philosophers and humanists of all kinds – not to mention theologians and others with a stake in the business – have written much on the subject. But what about a fresh approach? What about evolutionary biology? Can it, does it, say something on the subject and is that something of any real value? As it happens, since Darwin in the *Descent* – indeed, since before Darwin and going back at least to David Hume ([1757] 1963) – there has been interest in providing a natural account of religion and its development. In recent years, evolutionary biologists have been moving into this field in numbers. Yet the results hardly inspire confidence. Based on an understanding of religion that is (at the most charitable) sketchy, Dennett predictably looks upon religion as a meme of a rather unpleasant kind. It is a parasite – Dawkins (1997) calls it a "virus" – on humans, just as much as the liver fluke is a parasite on sheep. It serves its own ends without regard for the well-being of the host in which it resides. It leads Muslims, Christians, and Jews to devote "their lives to spreading the Word, making huge sacrifices, suffering bravely, risking their lives for an idea. So do Hindus and Buddhists" (Dennett 2006, 4). To be fair, sniffs Dennett, secular humanists are often not much better in these regards.

Others are a bit more optimistic. One who seems to think that religion is fairly heavily rooted in biology and is adaptive is the sociobiologist Edward O. Wilson. He writes:

> The highest forms of religious practice, when examined more closely, can
> be seen to confer biological advantage. Above all, they congeal identity.
> In the midst of the chaotic and potentially disorienting experiences
> each person undergoes daily, religion classifies him, provides him with
> unquestioned membership in a group claiming great powers, and by
> this means gives him a driving purpose in life compatible with his self
> interest. (E. Wilson 1978, 188)

Wilson recognizes that sometimes religions go a bit haywire biologically,
the Shakers for instance notoriously did not have sexual intercourse, but
regards them as anomalies, not likely to last long. Most religions put a
premium on reproduction. As it happens, Wilson does not spend much
time looking at the details of religious practice, trying to link them with
adaptive advantage. Others have tried to do so, for instance suggesting
that celibate orders associated with religions, like monks and nuns, often
provide help and status for close members of the family, thus leading to
increased reproduction there. Celibates are rather like worker ants, in fact
(Reynolds and Tanner 1983).

Not every Darwinian thinks religion adaptive. Hume thought religion a
by-product of other human features or habits. "We find human faces in the
moon, armies in the clouds; and by a natural propensity, if not corrected by
experience and reflection, ascribe malice or good-will to everything, that
hurts or pleases us" (Hume [1757] 1963, 41). Darwin himself argued along
the same lines, thinking that the "tendency in savages to imagine that
natural objects and agencies are animated by spiritual or living essences"
is illustrated by the mistaken actions of his dog (a beast, he assures us, who
is "a full-grown and very sensible animal"). Lying on the lawn, Darwin's
dog was perturbed by a parasol moving in the wind. Going on the attack
"every time that the parasol slightly moved, the dog growled fiercely and
barked. He must, I think, have reasoned to himself in a rapid and uncon-
scious manner, that movement without any apparent cause indicated the
presence of some strange living agent, and that no stranger had a right to
be on his territory" (C. Darwin 1871, I, 67).

Similar kinds of thinking can be found in today's evolutionary accounts
of religion. Student of culture Pascal Boyer (2002) inclines to think, with
Darwin, that religion in some way subverts features and habits that in
their own right are perfectly adaptive and functioning. "The building of

religious concepts requires mental systems and capacities that are there anyway, religious concepts or not. Religious morality uses moral intuitions, religious notions of supernatural agents recruit our intuitions about agency in general, and so on. This is why I said that religious concepts are parasitic upon other mental capacities" (311). Anthropologist Scott Atran (2004) backs up this kind of thinking by suggesting that religion perhaps subverts our selection-caused reactions to threatening events.

> Natural selection designs the agency-detection system to deal rapidly and economically with stimulus situations involving people and animals as wired to respond to fragmentary information under conditions of uncertainty, inciting perception of figures in the clouds, voices in the wind, lurking movements in the leaves, and emotions among interacting dots on a computer screen. This hair-triggering of the agency-detection mechanism readily lends itself to supernatural interpretation of uncertain or anxiety-provoking events. (78)

And so it goes on. There is more if you want to keep looking. Everyone seems to have their own angle. So, bringing this chapter to an end, do we conclude that all effort put into understanding human evolution is just a waste of time? Not at the physical level, obviously, where there is much positive activity and high-quality science, however judged. Nonetheless, picking up on the points made at the beginning of this section, one can start to see why people – including many philosophers – feel uncomfortable about the very program of trying to apply Darwinian evolutionary theory to human culture in any significant sense. Gould may be off-target with some of his general charges, but the meme "just so" stories critique does lodge at the back of one's mind. Discussions are simply all over the place. It is not so much a matter of ideology leading one to reject this stuff but that one is repelled by bad or casual or hopeful (or hopeless) attempts at doing science.

Others will probably make judgments that are not so harsh. To be fair, many evolutionary psychologists (using the term in the generic sense) and sympathizers would probably be the first to agree that thus far we have made little progress on many aspects of culture. They would argue that when it comes to a complex phenomenon like religion – which really covers a host of practices and beliefs, many of which are called "religious" only by courtesy or because we don't know where else to park them – we should not expect, at least at this stage of the game, to find unanimity or soundly grounded theories. We are still very much at the preliminary stage. However, they would also argue that there are areas within or overlapping with the domain of religion – particularly in understanding the nature of reasoning and of social behavior and morality – where much greater progress has been made. What is called for is not wholesale rejection but patient step-by-step attacks on the issues, combined no doubt with insights from other fields, something that happens far less than it should.

I am afraid therefore that there is no neat ending to this chapter. Clearly, taken as a whole, it would be ludicrous to deny that we have a functioning theory of human evolution. Both in fact and in theory there is successful and sophisticated understanding. But there is a long way to go, so far that some might wonder whether we have taken more than the first few steps, or will ever do so. As noted, we shall be returning to some of the pertinent issues in later chapters, so until then perhaps we should reserve our full judgment. For now, let us move the discussion forward by turning to a major issue that has been hinted at in previous chapters and that really is of vital concern when dealing with human evolution.

4   Progress

It is a huge job to understand humans from an evolutionary perspective, if indeed that is in any way possible. Especially since, even if you wanted to, for practical as well as ethical reasons, you cannot really experiment on us. But there is more to the matter than just the science, or rather just the straight science. There is the fact that we are humans. The conclusions matter in a very personal way. For Christians and indeed for members of other major faiths (especially Judaism and Islam) it is virtually a priori obvious that what evolution has to say about humankind is going to be interesting, and perhaps threatening. It is central to Christianity that we humans are special, we are particularly favored by God because we alone are "made in His image." What this means precisely is a matter of some debate. Most Christians are not biblical literalists, thinking that God created humans some 6,000 years ago, on the sixth day, out of mud or some such thing. They may not even think that we have to be exactly as we are. Perhaps for instance humans might have six fingers or green skin. I am not sure about sex, but perhaps even this is an option. It has always been fundamental to the Judeo-Christian perspective that God is not a sexual being, even though traditionally we refer to God as "Father" and as "He." But, as St. Augustine pointed out around the end of the fourth century, we do have to be intelligent beings and creatures with a moral capacity. Without that, nothing makes sense. More than this, our existence has to be necessary in some sense. It cannot just be a matter of chance that humans or human-like creatures exist. I don't see any reason why we shouldn't exist on the moon or somewhere else if that is a viable option. (Obviously as things stand the moon is not a viable option.) But exist we must (Plantinga 2009).

Non-believers obviously don't have this kind of motivation to make humans special and essential, as we might say. Yet a surprisingly large number do share the sentiments to some extent. A large factor in the opposition to evolutionary psychology and related activities is more theological (or, if you like, philosophical) than the raw facts of the case would lead you to expect. People really just don't like the idea that we are simply part of the animal world, no more, no less. (See Fodor and Piattelli-Palmarini 2010, for example.) Somehow this is seen as challenging to our status, and something that downgrades us unjustifiably. This sentiment is perhaps understandable from those who are Marxists, for not only is this (via Hegel) a belief-system with deeper roots in Christianity than its adherents are often prepared to recognize, but explicitly it singles out human beings as entities uniquely in the realm of culture, tool use, and so forth. But the feeling is broader than that. The twentieth-century American paleontologist George Gaylord Simpson (1949) spoke for many when he said: "If animals are so important, why don't they just say so?"

Evolution as an epiphenomenon of Progress

Without taking a stand on these feelings, let us dig into the question. What does evolutionary biology, Darwinian evolutionary biology in particular, have to say about human uniqueness and our inevitable appearance? Start with history, for it tells us much about the case (Ruse 1996). Above all, it tells us that the concern with the special status of humans goes back to the beginnings of evolutionary hypothesizing. We have seen that it was not really until the beginning of the eighteenth century that people first started to speculate on natural developmental origins for life. What we should understand now is that these speculations did not occur in a vacuum, but rather were the direct outcome of a metaphysical or theological battle about the nature and status of humankind (Ruse 2005b). I have explained earlier that although there were indeed those who took the Bible more or less literally, it has always been part of Christian tradition (articulated by St. Augustine) that if need be one can understand the text metaphorically or allegorically in order to harmonize it with other beliefs, from science and philosophy and so forth. At the beginning of the eighteenth century, the all-important belief for the Christian was Providence. We are the creatures of God, we have fallen into sin, and the only hope of salvation is through the Cross and God's grace, his undeserved love and forgiveness. To quote the words of the great hymn by the Congregationalist minister Isaac Watts:

> When I survey the wondrous Cross
> On which the Prince of Glory died
> My richest gain I count but loss
> And pour contempt on all my pride.

Also explained earlier was the opposition to all of this by more secular thinkers. They embraced the ideology or metaphysic of Progress, the belief that through our intelligence and labors we humans can make things better, in science, in education, in healthcare, and more. Evolution was part and parcel of this ideology. People read Progress into the organic world, seeing it as an upward ladder from the primitive to the complex, from what they called the "monad" to what they called the "man." This may have been a bit of a stretch, but it was not such an odd thing to do, because, even though there was no proper fossil record or anything like that, there was a tradition going back to the Greeks of seeing the living world as connected in a series from least to most, the so-called "great chain of being" (Lovejoy 1936). Having done this, they promptly turned around and saw the Progress of the living world as justification for their beliefs in social Progress! Charles Darwin's grandfather Erasmus has been mentioned, but he was far from alone. The French *encyclopédiste* Denis Diderot is a good example. "Just as in the animal and vegetable kingdoms, an individual begins, so to speak, grows, subsists, decays and passes away, could it not be the same with the whole species?" (Diderot 1943, 48, quoting *On the Interpretation of Nature*, 1754). He made no bones about being a Progressionist and seeing a link between his social views and his scientific speculations. "The Tahitian is at a primary stage in the development of the world, the European is at its old age. The interval separating us is greater than that between the new-born child and the decrepit old man" (Diderot 1943, 152, quoting *Supplement to Bougainville's Voyage*, 1772).

Progress-infused thoughts continued unabated in evolutionary speculations right down to the middle of the nineteenth century. In fact, it was often the very possibility of promoting social and cultural Progress through talk about the fossils and today's organisms that motivated people to get involved in such speculations. Typical was the then-unidentified author of *The Vestiges of the Natural History of Creation* published in 1844 (some fifteen years before the *Origin*, but several years after Charles Darwin had become an evolutionist and discovered natural selection). We now know the author to have been Robert Chambers, a leading Scottish businessman, who with his brother had industrialized the printing industry and was a publisher of widely read books and magazines. He was deeply committed to the upward march of society, which came right out in the *Vestiges*:

> A progression resembling development may be traced in human nature, both in the individual and in large groups of men … Now all of this is in conformity with what we have seen of the progress of organic creation. It seems but the minute hand of a watch, of which the hour hand is the transition from species to species. Knowing what we do of that latter transition, the possibility of a decided and general retrogression of the highest species towards a meaner type is scarce admissible, but a forward movement seems anything but unlikely. (Chambers 1846, 400–402)

Darwin and Progress

We come now to Charles Darwin and the *Origin of Species*. With some good reason, many people think that Darwin rang the death knell on thoughts of biological progress, or if he did not do the job fully he started the process that finished with the coming of modern genetics. Natural selection may not be a tautology, but it does relativize change. It says that the fittest survive and reproduce. But for all that there is a uniformity claimed about the fittest – what is fit in one case is fit in similar cases – there is non-uniformity when (as is usually the case) things are not similar. Is it better to be big and strong, or small and possibly weak? Well, it all depends. If there is lots of food which takes little time to find and gather, then big and strong is probably the better strategy. But (as possibly in the case of the hobbit), if food supplies are limited, you might be better off going for the small option, especially if predators are uncommon or rare. It is the same with other features, even intelligence. Having big brains requires lots of protein and hence probably, since big protein supplies generally come packaged in the form of animals, lots of effort to get it. It does not necessarily follow that the effort is worth it. If there is lots of cheap fodder around like grass, you might be better off going vegetarian and spending your time grazing and munching. In the immortal words of the late Jack Sepkoski, a paleontologist: "I see intelligence as just one of a variety of adaptations among tetrapods for survival. Running fast in a herd while being as dumb as shit, I think, is a very good adaptation for survival" (Ruse 1996, 486).

The coming of Mendelian genetics is often thought to have finished what Darwin started. Darwin himself was always insistent that the new variations that appear in populations, the raw building blocks of evolution, are random – not in the sense of uncaused, although he did not know the causes (he speculated that injury to the generative organs might play a role) he was sure there were causes – but in the sense of not appearing according to need. To suppose that they did appear as needed seemed, to Darwin, to put in a directedness, a teleological impulse, that is contrary to modern science. He was very critical of his American supporter Asa Gray when the latter did suppose that the variations of evolution have some kind of direction. For Darwin, this was to take the whole matter outside of science (Ruse 1979). And the arrival of modern genetics underscored this very point. The whole point about the new variations, what we now call "mutations," is that they do not occur according to need. Today, we know a lot about the causes and we know that the rate of mutation can usually be quantified, but individual mutations are random both in their not appearing just when needed and even more in being unrelated to need. An organism may be able to use them, but it is a crap shoot all of the way. So once again it seems that thoughts of overall direction to the course of Darwinian (or neo-Darwinian) evolution are stymied.

So much for theory. In fact the practice, or rather the history, was very different. Charles Darwin always believed in social and cultural Progress. Why would he not? He was a child of the Industrial Revolution, a Briton living at a time when his country ruled the waves and much of the land, too. It permeated his thinking and most of that of his countrymen. Consequently, after the *Origin*, people – including Charles Darwin himself – went right on being enthusiastic progressionists, in biology that is. "The inhabitants of each successive period in the world's history have beaten their predecessors in the race for life, and are, in so far, higher in the scale of nature; and this may account for that vague yet ill-defined sentiment, felt by many palæontologists, that organisation on the whole has progressed" (C. Darwin 1859, 345). By the end of the book, it is clear that in the mind of Charles Darwin, the sentiment is not so very vague. "And as natural selection works solely by and for the good of each being, all corporeal and mental endowments will tend to progress towards perfection" (C. Darwin 1859, 489). Or consider the closing lines of the *Origin*.

> From the war of nature, from famine and death, the most exalted object which we are capable of conceiving, namely, the production of the higher animals, directly follows. There is a grandeur in this view of life, with its several powers, having been originally breathed into a few forms or into one; and that, whilst this planet has gone cycling on according to the fixed law of gravity, from so simple a beginning endless forms most beautiful and most wonderful have been, and are being, evolved. (490)

In many respects, the *Descent of Man* is a more popular book than the *Origin* – it is packed with anecdotes and is well larded with Darwin's own opinions and value judgments. Among which are the beliefs that we humans are at the top of the heap and among humans the inhabitants (or former inhabitants) of Europe are the very best. A neat combination of an endorsement of capitalism and the virtues of progress in culture gets the desired result. "In all civilised countries man accumulates property and bequeaths it to his children. So that the children in the same country do not by any means start fair in the race for success. But this is far from an unmixed evil; for without the accumulation of capital the arts could not progress; and it is chiefly through their power that the civilised races have extended, and are now everywhere extending, their range, so as to take the place of the lower races" (C. Darwin 1871, I, 169).

Had nothing changed with the coming of the theory of evolution through natural selection? Actually, a little did change. Although Darwin may have been as much of a biological progressionist as earlier evolutionists, he showed his greater degree of scientific sophistication in realizing that you cannot just talk of progress without at some level trying to define it, or rather the things that make for progress. Everyone knows that what you are after is evolution up to beings that have properties that coincide with "human-like," but to avoid circularity or triviality you have got to have some kind of independent criteria. If you define biological progress in terms of "human-like" then obviously you are going to get progress in some sense. We humans are still here, and we are the final product of evolution (or one of the final products). But this is not much of a conclusion. You have got to find criteria separate from simply "human-like" – you

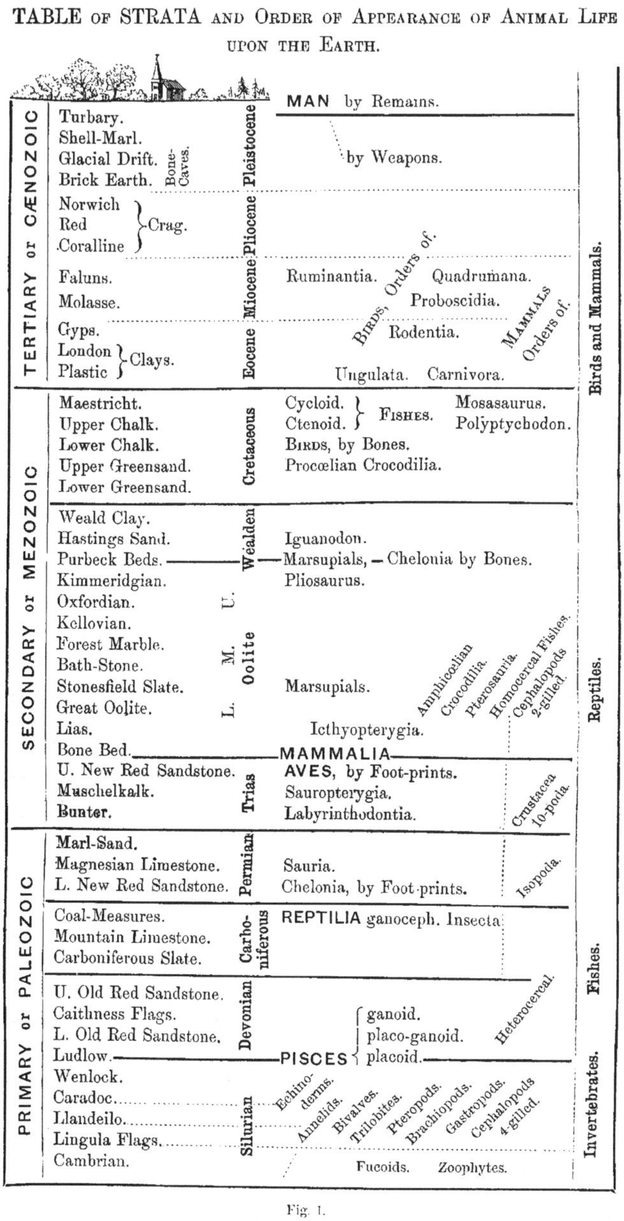

TABLE OF STRATA AND ORDER OF APPEARANCE OF ANIMAL LIFE
UPON THE EARTH.

TERTIARY or CÆNOZOIC	Turbary. Shell-Marl. Glacial Drift. Brick Earth. } Bone-Caves.	Pleistocene	MAN by Remains.
			by Weapons.
	Norwich } Red } -Crag. Coralline }	Pliocene	
	Faluns. Molasse.	Miocene	Ruminantia. Orders of. Quadrumana. Proboscidia.
	Gyps. London } -Clays. Plastic }	Eocene	Rodentia. MAMMALS Orders of. Ungulata. Carnivora.

Birds and Mammals.

SECONDARY or MEZOZOIC	Maestricht. Upper Chalk. Lower Chalk. Upper Greensand. Lower Greensand.	Cretaceous	Cycloid. } Ctenoid. } FISHES. Mosasaurus. Polyptychodon. BIRDS, by Bones. Proccelian Crocodilia.
	Weald Clay. Hastings Sand. Purbeck Beds. Kimmeridgian. Oxfordian. Kellovian. Forest Marble. Bath-Stone. Stonesfield Slate. Great Oolite. Lias.	Wealden / U. M. Oolite / L.	Iguanodon. Marsupials, — Chelonia by Bones. Pliosaurus. Marsupials. Icthyopterygia.
	Bone Bed. U. New Red Sandstone. Muschelkalk. Bunter.	Trias	MAMMALIA AVES, by Foot-prints. Sauropterygia. Labyrinthodontia.

Reptiles.

Amphicoelian / Crocodilia / Pterosauria / Homocoercal Fishes / Cephalopods 2-gilled.

Crustacea 10-poda.

PRIMARY or PALEOZOIC	Marl-Sand. Magnesian Limestone. L. New Red Sandstone.	Permian	Sauria. Chelonia, by Foot-prints.
	Coal-Measures. Mountain Limestone. Carboniferous Slate.	Carboniferous	REPTILIA ganoceph. Insecta
	U. Old Red Sandstone. Caithness Flags. L. Old Red Sandstone. Ludlow.	Devonian	{ ganoid. placo-ganoid. PISCES { placoid.
	Wenlock. Caradoc. Llandeilo. Lingula Flags. Cambrian.	Silurian	Echinoderms. Annelids. Bivalves. Trilobites. Pteropods. Brachiopods. Gastropods. Cephalopods 4-gilled. Fucoids. Zoophytes.

Isopoda. / Heterocercal. / Fishes. / Invertebrates.

Fig. 1.

Figure 4.1 The fossil record as known when the *Origin* was published. From Owen (1861).

have got to show that "human-like" emerges independently. Relying on the best recent German biology, Darwin (writing in the third edition of the *Origin* in 1861) opted for some kind of organization and differentiation and specialization.

Natural selection acts, as we have seen, exclusively by the preservation
and accumulation of variations, which are beneficial under the
organic and inorganic conditions of life to which each creature is at
each successive period exposed. The ultimate result will be that each
creature will tend to become more and more improved in relation to
its conditions of life. This improvement will, I think, inevitably lead to
the gradual advancement of the organisation of the greater number of
living beings throughout the world. But here we enter on a very intricate
subject, for naturalists have not defined to each other's satisfaction what
is meant by an advance in organisation. Amongst the vertebrata the
degree of intellect and an approach in structure to man clearly come
into play. It might be thought that the amount of change which the
various parts and organs undergo in their development from the embryo
to maturity would suffice as a standard of comparison; but there are
cases, as with certain parasitic crustaceans, in which several parts of
the structure become less perfect, so that the mature animal cannot be
called higher than its larva. Von Baer's standard seems the most widely
applicable and the best, namely, the amount of differentiation of the
different parts (in the adult state, as I should be inclined to add) and
their specialisation for different functions; or, as Milne Edwards would
express it, the completeness of the division of physiological labour. (C.
Darwin 1861, 133)

How does this tie in with evolution, or more precisely, how does this tie in
with natural selection and human nature? Darwin was acutely sensitive to
the fact that there was an issue here. He could not simply expect progress to
emerge. Even more importantly, although he was happy to accept German
thinking to define progress, he did not want to get identified with the kind
of Germanic upward P/progressivism one finds in German Romantic philosophers (like Fichte and Schelling) at the beginning of the nineteenth century – a sort of world force that was supposed by Hegel to be pushing the
whole of life ever upward (Richards 2003). It was for this reason that on
the flyleaf of his own copy of *Vestiges* Darwin wrote "never use higher and
lower." It was not the Progressivist sentiment that he repudiated, but the
belief (clearly endorsed by Chambers) that life has a kind of upward momentum, all of its own. Natural selection was the mechanism for Darwin, and
he recognized that sometimes it stands still or goes backward, and we cannot expect instant or constant biological progress. Often, indeed, even using
specialization as your measure, it is hard to know what to think:

[W]e shall see how obscure a subject this is if we look, for instance, to fish, amongst which some naturalists rank those as highest which, like the sharks, approach nearest to reptiles; whilst other naturalists rank the common bony or teleostean fishes as the highest, inasmuch as they are most strictly fish-like, and differ most from the other vertebrate classes. Still more plainly we see the obscurity of the subject by turning to plants, with which the standard of intellect is of course quite excluded; and here some botanists rank those plants as highest which have every organ, as sepals, petals, stamens, and pistils, fully developed in each flower; whereas other botanists, probably with more truth, look at the plants which have their several organs much modified and somewhat reduced in number as being of the highest rank. (C. Darwin 1861, 133–34)

Nevertheless, as intimated in a passage given above, ultimately Darwin thought that natural selection really does make for upward change, because the winners overall will be better than the losers. And this ties in with the definition of progress.

If we look at the differentiation and specialisation of the several organs of each being when adult (and this will include the advancement of the brain for intellectual purposes) as the best standard of highness of organisation, natural selection clearly leads towards highness; for all physiologists admit that the specialisation of organs, inasmuch as they perform in this state their functions better, is an advantage to each being; and hence the accumulation of variations tending towards specialisation is within the scope of natural selection. (C. Darwin 1861, 134)

After Darwin

Leave critical comment for the moment, for we shall be coming back to the modern-day equivalent of Darwin's own thinking. Accept simply that even if Darwin worried about progress, few others did. Almost to a person, people went on thinking of evolution as progressive. Extreme but highly influential was Darwin's contemporary Herbert Spencer. Here he is in a passage written two years before the *Origin*, but to be repeated by him non-stop for the rest of his very long life (into the twentieth century). Like Darwin in being influenced by Germanic thinking, Spencer also adopted a criterion of Progress that involved division and specialization, or as he called it a move from the homogeneous to the heterogeneous:

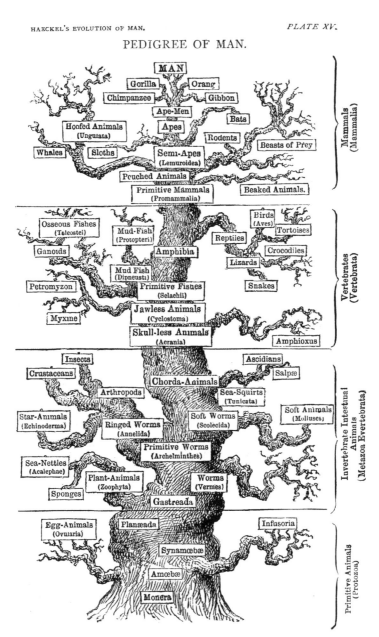

Figure 4.2 The tree of life as drawn by Darwin's German supporter Ernst Haeckel (1896). Note how humans come at the top of a progressive growth upward.

Now we propose in the first place to show, that this law of organic progress is the law of all progress. Whether it be in the development of the Earth, in the development of Life upon its surface, in the development of Society, of Government, of Manufactures, of Commerce, of Language, Literature, Science, Art, this same evolution of the simple into the complex, through successive differentiations, holds throughout. From the earliest traceable cosmical changes down to the latest results of civilization, we shall find that the transformation of the homogeneous into the heterogeneous is that in which Progress essentially consists. (H. Spencer 1857, 246)

In Spencer's eyes, everything obeys this law. Compared with other animals, humans are more complex or heterogeneous; compared with savages, Europeans more complex or heterogeneous; and compared with the tongues of other peoples, the English language more complex or heterogeneous. One should say that there was a certain inevitability about all of this that was alien to Darwinism. Nor was there any striving to fit this thinking to more selection-based notions like the division of labor. Although he had thought of natural selection before the *Origin* was published, for Spencer it was never a major part of the evolutionary story. A kind of Germanic progressivist determinism was there for all to see. Causes just keep multiplying effects and this leads to ever-greater complexity, and somehow value or worth gets pulled in along the way. This all seems to occur in waves. Something disturbs the natural balance, forces work to regain the balance, but when this occurs we have moved a stage higher. An overall process of "dynamic equilibrium."

What about the coming of Mendelism? Take two of the key figures bringing together Darwinian selection and Mendelian genetics: Ronald A. Fisher in Britain and Sewall Wright in America. Both were ardent Progressionists! Fisher (1930) made much of something that he called the "Fundamental theorem of evolution." There is a kind of perpetual upward momentum to the evolutionary process, as the result of selection at work. "The rate of increase in fitness of any organism at any time is equal to its genetic variance in fitness at that time" (35). In other words things are getting better all of the time, proportional to the amount of variation in a population. This does not mean that you can expect constant upward change. It is all a bit like the stock market in that respect. "Against the action of natural selection in constantly increasing the fitness in every organism … is to be set off the very considerable item of the deterioration of its organic and inorganic environment" (42). Overall, though, things are moving upward – even if Fisher was very worried about the effects of civilization on humans. Unfortunately it appears that one major consequence of culture is that the best and brightest do not have as many offspring as those lower down the scale, and this implies the degeneration of the human species. To alleviate this, Fisher became an ardent eugenicist, arguing that we need to encourage the upper levels of society to have more children. Whether his prescriptions have any force will be a topic for a later chapter. Here we can note that he himself married a young woman for her breeding potential and then proceeded to have eight children!

As we saw in chapter 1, although they agreed on the formal mathematics, Sewall Wright took a very different path from Fisher. We saw there also the elements that make very easy the case for Wright's enthusiasm for all kinds of P/progress, in culture and in biology (Ruse 2004). Much influenced by his work at the United States Department of Agriculture on the genetics of cattle breeding, and even more influenced by an enthusiasm for the ideas of Herbert Spencer, Wright saw the key to innovative evolutionary change in the breaking up of large populations into small groups in which genetic drift would be effective, and then recombination into the larger mass once again. He thought that in the small groups new innovative adaptations would appear thanks to drift and that these would then spread in the reunited whole. Combining this with the metaphor of an adaptive landscape, Wright saw organisms as moving from one peak to another thanks to the new adaptations. One could of course regard the overall landscape as akin to a water bed, with a new peak in one area being balanced by a new valley in another, and with no overall upward gain in height. However, this was decidedly not Wright's own vision, as he saw organisms inching step by step toward greater heights – "dynamic equilibrium" – ultimately perhaps into one universal world soul. Parenthetically one might add that Wright had some very odd metaphysical views about the relationships between body and mind. He was not the first or the last to combine an ability to perform abstruse mathematics with a taste for wild philosophical systems of a quasi-mystical kind.

Biological progress was alive and well in the 1930s. The theoreticians gave the lead and the empiricists were happy to follow. There is good evidence that this was a major attraction for the leading researchers in the field – Dobzhansky, Mayr, and Simpson in America and overwhelmingly Julian Huxley (the grandson of T. H. Huxley) in Britain. For the last named, his whole intellectual life was a mission to replace conventional Christianity with what (in the title of one popular book) he was to call "religion without revelation." Deeply influenced by the thinking of the French vitalist philosopher Henri Bergson (1907), Huxley argued that the key to understanding lies in control and independence – basically the ability to manipulate one's environment and to break loose from its constraints. Thus judged, humans are clearly the winners in the race to the top. Curiously for one who for all of his very long life carried high the banner of Darwinism, or more precisely who inherited the role of top Darwinian, Julian Huxley was always a bit ambivalent about the workings and power of natural selection. Perhaps given his grandfather's like ambivalence, his stance was not so curious after all. On the one hand, he did think that selection had a hand in biological progress. "It should be clear that if natural selection can account for adaptation and for long-range trends of specialization, it can account for progress too. Progressive changes have obviously given their owners advantages which have enabled them to become dominant "(J. Huxley 1942, 568). On the other hand, like Bergson (who was in turn influenced by Spencer), he saw an inevitability to the upward rise of humankind. "One somewhat curious fact emerges from a survey of biological progress as culminating for the evolutionary moment in the dominance of *Homo sapiens*. It could apparently have pursued no other general course than that which it has historically followed" (569).

Biological progress today

Move on now from history to the present. My suspicion is that, with good reason, many if not most evolutionary biologists today would say that all of this talk about biological progress really is past history. They would say that today no reputable evolutionary biologist, certainly no Darwinian biologist, believes in biological progress. Certainly we do not find discussions of it in the literature, in the major journals like *Evolution* and *American Naturalist*. This last point is certainly true, but the story is a little more complex than that, and history tells us why. (It is curious how few scientists think that history can tell them anything. This includes evolutionary biologists, who, as a matter of professional commitment, believe that the biological past can tell us virtually everything.) The reason why biological progress fell out of favor in professional evolutionary biological circles was not that people suddenly stopped believing in it. It was because increasingly they realized that parading so obviously a value-laden notion as progress right up front in their science was antithetical to their aims as successful professional scientists. From the Scientific Revolution on there was an ideology, particularly in the physical sciences, that science should in some sense be purely objective knowledge. In the felicitous phrase of Karl Popper (1972), it should be "knowledge without a knower" – not, obviously, in the sense that there is no knower involved, but that the nature of the knower (male, female, Christian, Jew, British, American, gay, straight) should be entirely irrelevant. For various reasons – not the least the rise of Nazi Germany and its condemnation of "Jewish science" – this ideology burned fiercely in the 1930s and 1940s.

Yet here was evolutionary biology, just now rising from many years of (at best) museum-level displays and wild hypothesizing, still parading front and center beliefs about human superiority – often still parading views about white human superiority. If this was not enough, there were other branches of the biological sciences, most notably the growing molecular branches, that were competing for students and grants and university places and so forth. So a deliberate effort was made to downplay and remove talk of biological progress from the professional publications. By the 1950s, as you search the literature, it is harder and harder to find discussions of upward rise anywhere. The one exception was Julian Huxley, who went on promoting biological progress to the end. But it was long since he had held a university position, and increasingly he was professionally isolated from mainstream science. When, in the late 1950s, Huxley (1959) endorsed the ideas of the French paleontologist Pierre Teilhard de Chardin SJ (1955), who was promoting a progressionist view of life leading up to humans (and beyond to Christ), the reactions were scathing (Medawar 1967). It was not so much that Julian Huxley was wrong, but that he was openly doing things that his community saw as flagrantly opposed to the status-building in which they were engaged.

What does this mean for today? The progress-expelling campaign was obviously successful, and one suspects that a major effect has been that today's professional evolutionist has moved on. He or she is not drawn to the field because they have Progressivist yearnings, and for them keeping biological progress out of the discussion is not something that has to be done deliberately. The wonder would be that one might want to bring it in. However, having said this, since the main motivation for expelling biological progress was more social than intellectual, one might expect to find that there will still be evolutionists today who have sympathies of one sort or another with progress in some sense, and that these will come out, if not in the fully professional writings then in the more popular (or philosophical) writings about evolution and its history.

And this is indeed so. It does not take much searching to find thoughts of biological progress and claims that link it to modern evolutionary biology. Today's most distinguished living evolutionist, Edward O. Wilson of Harvard University, is open in his fervent belief in biological progress. "The overall average across the history of life has moved from the simple and few to the more complex and numerous. During the past billion years, animals as a whole evolved upward in body size, feeding and defensive techniques, brain and behavioral complexity, social organization, and precision of environmental control – in each case farther from the nonliving state than their simpler antecedents did" (E. Wilson 1992, 187). Adding: "Progress, then, is a property of the evolution of life as a whole by almost any conceivable intuitive standard, including the acquisition of goals and intentions in the behavior of animals." With views like this coming down from on high (aka Harvard University), it is little surprise that in the general domain also biological progress is alive and well. In the popular mind, in the way that evolution is presented in popular print or on radio or television, the course of biological history is almost inevitably represented as one leading in a fairly straight line up to *Homo sapiens*. To the person in the street, it would come as a significant surprise to discover that this is not the universal commitment of all practicing evolutionary biologists.

Reasons for biological progress

Recognizing then that biological progress will be something that yields beings human-like (although at the risk of circularity it cannot make direct reference to being human-like), and reserving for a moment discussion of how this might be done, let us look at some recent suggestions for achieving

Figure 4.3 The almost inevitable popular picture of human evolution, progressive through and through.

such progress. Some, I think, are perhaps of broader interest but can be ruled out quickly as suggestions meriting scientific status or consideration. Going back briefly to Christianity (and other faith systems) the simple fact is that it cannot allow the possibility of failure. For the Christian, humans or human-like beings had to emerge. That is not an option that can fail (Ruse 2001, 2010). The obvious solution therefore is to put God himself up front doing the job. That was what Asa Gray (1876) and other so-called "theistic evolutionists" always did. They made God responsible for the new variations, the mutations in today's language, and He directed enough of them that humans were bound to emerge. Today's chief enthusiast for this kind of thinking is the American physicist-theologian Robert John Russell (2008), who thinks that God works down at the quantum level, directing changes as needed. As far as we are concerned, His actions are hidden because we can see only the average effects, but He makes sure that at the individual level of change the right variation appears at the right time.

Hidden or not, this is not science, and there are those who think it is not very good Christianity either. If God does work in this sort of way, why (if He is an all-powerful, loving Father) does He not also work to prevent some of the horrendous mutations that cause untold human suffering? The trouble with getting God involved in the day-to-day workings of evolution is that, although He may get credit for the good things, He makes himself directly liable for the bad things, also. So let us move on to suggestions that, be they true or false, are at least putatively scientific. There are at least three suggestions, although as we shall see, the third is perhaps better considered an alternative to progressivist readings of the evolutionary record.

The first, endorsed most enthusiastically today by Richard Dawkins (1986), stands in direct line back to Darwin. It picks up on his belief that selection leads to competition, with one group or line outcompeting another group or line, and that this leads bit by bit to overall improvement (Dawkins and Krebs 1979). Humans emerge as the end of the line. The metaphor of an arms race was unknown in the middle of the nineteenth century (although Asa Gray in a letter to Darwin does use the word "race" in this context) but it is this that has framed the discussion in the twentieth. It was Julian Huxley who first and most explicitly spelt out this kind of thinking. It comes in a little neo-Bergsonian book he published in 1912. He gave a graphic description of an arms race couched in terms of the then-state-of-the-art naval military technology. "The leaden plum-puddings were not unfairly matched against the wooden walls of Nelson's day." Now, however, "though our guns can hurl a third of a ton of sharp-nosed steel with dynamite entrails for a dozen miles, yet they are confronted with twelve-inch armor of backed and hardened steel, water-tight compartments, and targets moving thirty miles an hour. Each advance in attack has brought forth, as if by magic, a corresponding advance in defence." Likewise in nature, "if one species happens to vary in the direction of greater independence, the inter-related equilibrium is upset, and cannot be restored until a number of competing species have either given way to the increased pressure and become extinct, or else have answered pressure with pressure, and kept the first species in its place by themselves too discovering means of adding to their independence" (J. Huxley 1912, 115–16). Eventually: "it comes to pass that the continuous change which is passing that through the organic world appears as a succession of phases of equilibrium, each one on a higher average plane of independence than the one before, and each inevitably calling up and giving place to one still higher." Not just Bergson but Herbert Spencer seems to be the inspiration behind this last passage, and this perhaps accounts for the inevitability that one senses in Huxley's thinking on the topic.

Richard Dawkins may not share Huxley's enthusiasm for Spencer and Bergson, but he is very much in favor of progress. "Directionalist common sense surely wins on the very long time scale: once there was only blue-green slime and now there are sharp-eyed metazoa" (Dawkins and Krebs 1979, 508). And arms races are the key. In particular, using today's refinements, he points out that, more and more, arms races rely on computer technology rather than brute power, and – in the animal world – Dawkins translates this into bigger and bigger brains. No prizes are given for guessing who has won. He refers to a notion known as an animal's EQ, standing for "encephalization quotient" (Jerison 1973). This notion is a kind of cross-species measure of IQ that factors out the amount of brain power needed simply to get an organism to function (whales require much bigger brains than shrews because they need more computing power to get their bigger bodies to function), and that then scales according to the surplus left over. Dawkins (1986) writes:

> The fact that humans have an EQ of 7 and hippos an EQ of 0.3 may not literally mean that humans are 23 times as clever as hippos! But the EQ as measured is probably telling us *something* about how much "computing power" an animal probably has in its head, over and above the irreducible amount of computing power needed for the routine running of its large or small body. (189)

Even an organism with a low EQ probably does not need much help in making out the precise nature and import of that "something."

The second major attempt to articulate and justify a Darwinian basis for a progressivist rise in life, from the blob to the human, comes from the work of the British paleontologist Simon Conway Morris (2003). He starts his case from the fact that only certain areas of potential morphological space will be able to support functional life. As a Darwinian he adds to this the assumption that selection is forever pressing organisms to look for such potential, functional spaces. From this he draws the conclusion that, if such spaces exist, in the full course of time they will be occupied. For better or for worse, Conway Morris seems to think that the occupation will probably occur sooner rather than later, and probably many times. Now comes the key step. He highlights the way in which life's history shows an incredible number of instances of convergence – meaning by this cases where the same adaptive morphological space has been occupied again and again. There are many outstanding examples, among the most dramatic being that of saber-toothed, tiger-like organisms, where the

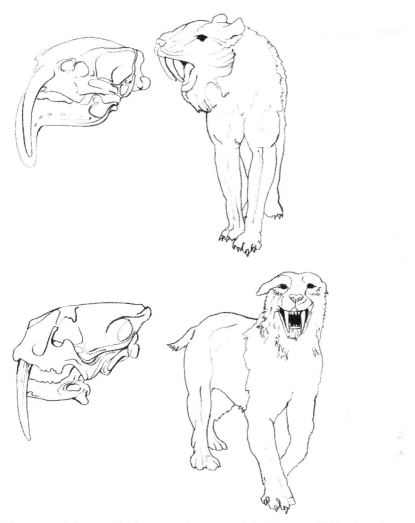

Figure 4.4 Saber-toothed mammals: marsupial thylacosmilid (top) and placental cat (bottom).

North American placental mammals (real cats) were matched right down the line by South American marsupials (thylacosmilids). There existed a niche for organisms that were predators, with cat-like abilities and shearing/stabbing-like capacity. Darwinian selection found more than one way to enter it. Indeed, Conway Morris points out, it may even have been that, long before the mammals, the dinosaurs likewise found this niche.

The overall conclusion that Conway Morris draws is that, because convergence is almost the norm rather than the exception, we must allow that the historical course of nature is not random but strongly selection-constrained along certain pathways and to certain destinations. Most particularly, movement up the order of nature, the chain of being, is no accident. For all of the contingency in the Darwinian evolutionary process, such a progress was predestined. Sooner or later therefore some kind of intelligent being (often called a "humanoid") was bound to emerge. After all, our own very existence shows that a kind of cultural adaptive niche exists – a niche that prizes intelligence and social abilities. Nor is this niche something with existence only in the context of humankind. There is reason to think it existed independently of us. Many other organisms have (with greater or lesser success) aspired to enter it. After all, we humans came to our niche adaptive step by adaptive step. First came things like eyes and ears and noses that allowed us to move up, niche by niche; then came the growth of brain power that pushed our ancestors (and co-evolvers) up to ever-new and empty niches; and with these improved brains and consequent greater intelligences came more sophisticated patterns of social behavior. In the end, came humankind, less by chance and more by Darwinian destiny.

> If brains can get big independently and provide a neural machine
> capable of handling a highly complex environment, then perhaps there
> are other parallels, other convergences that drive some groups towards
> complexity. Could the story of sensory perception be one clue that, given
> time, evolution will inevitably lead not only to the emergence of such
> properties as intelligence, but also to other complexities, such as, say,
> agriculture and culture, that we tend to regard as the prerogative of
> the human? We may be unique, but paradoxically those properties that
> define our uniqueness can still be inherent in the evolutionary process. In
> other words, if we humans had not evolved then something more-or-less
> identical would have emerged sooner or later. (Conway Morris 2003, 196)

The third proposal is, as I have said, less a proposal for progress per se than an alternative solution. Stephen Jay Gould was always very hostile to notions of biological progress. He was forever fulminating against it, because he thought it was racist and sexist and much more. Particularly (and at an obvious personal level) he disliked the way that progressivist ideas had been used to belittle the status and talents of Jews (Gould 1981). He spoke of biological progress as "a noxious, culturally embedded, untestable, nonoperational, intractable idea that must be replaced if we wish to understand the patterns of history" (Gould 1988, 319). He argued that there is nothing inevitable about the emergence of humans. Making joking reference to the asteroid that hit the Earth 65 million years ago and wiped out the dinosaurs, making possible the age of mammals, he wrote:

> Since dinosaurs were not moving toward markedly larger brains, and since such a prospect may lie outside the capabilities of reptilian design (Jerison, 1973; Hopson, 1977), we must assume that consciousness would not have evolved on our planet if a cosmic catastrophe had not claimed the dinosaurs as victims. In an entirely literal sense, we owe our existence, as large and reasoning mammals, to our lucky stars. (Gould 1989, 318)

And yet Gould did not want to deny that there may be something to the idea of an increase in life's complexity. Although unacknowledged, he used an idea anticipated (at the very least) in an early jotting by Charles Darwin himself. Somehow complexity just emerges because in a way you are always going to be building on what you have already.

> The enormous *number* of animals in the world depends of their varied structure & complexity. – hence as the forms became complicated, they opened fresh means of adding to their complexity. – but yet there is no **necessary** tendency in the simple animals to become complicated although all perhaps will have done so from the new relations caused by the advancing complexity of others. – It may be said, why should there not be at any time as many species tending to dis-development (some probably always have done so, as the simplest fish), my answer is because, if we begin with the simplest forms & suppose them to have changed, their very changes ton tend to give rise to others. (Notebook entry, January 1839, E 95–E 97, in Barrett *et al.* 1987, 422–23)

Darwin certainly thought of the complexity as ultimately adaptive. Immediately after the passage just quoted, he added: "it is quite clear that a large part of the complexity of structure is adaptation." I am not sure that Gould felt this way, but he did see complexity as growing during the course of life's history (Gould 1996). Life is a bit asymmetrical. Necessarily, it started simple. Necessarily, it cannot get less simple. However, it can get more complex. Not through any guiding power, but because this is the way that things are. It is a rather like the old tale of the drunkard and the sidewalk. On one side, the sidewalk is bounded by a wall, and on the other side lies the gutter. It may take a long time, but in the end the drunkard will end in the gutter. This is not through any conscious choice, but because the drunkard can fall off the sidewalk and he cannot walk through the wall. His random staggering will eventually lead to the gutter. So it is with evolution. There is no progress in nature, but there is direction. Will this direction eventually end up with humans or human-like beings? Interestingly, although he was not optimistic about this one-off world of ours, he rather thought that if you extend your gaze out to the rest of the universe and to the many life-forms that surely exist out there, one might reasonably expect to find the evolution of intelligent beings of some sort (Dick 1996).

Similar ideas have been promoted recently by paleontologist Daniel McShea and philosopher Robert Brandon (2010). They see a kind of non-Darwinian upward momentum to life's history. Introducing what they call the "zero-force evolutionary law" (ZFEL for short) they write: "In any evolutionary system in which there is variation and heredity, in the absence of natural selection, other forces, and constraints acting on diversity or complexity, diversity and complexity will increase on average" (3). Interestingly they circle back to Spencer, acknowledging that they write in his spirit. It seems that for them (as was the case for Spencer) things just naturally keep complexifying – one cause leads to several effects and these in turn multiply. They are not committed to the kind of surging view that characterizes dynamic equilibrium – although Gould's non-Darwinian punctuated equilibrium surely echoes – but they do see an upward drive as part of the ontology of the universe.

Is there biological progress?

In concluding this chapter, let us turn now to some critical thoughts. First, what about the very notion of progress itself? What about higher and lower and what about features that do increase or improve through life's history and that characterize human beings above all others? As soon as you start to look at it carefully, complexity begins to be an awfully slippery concept. Dawkins (1992, 2003) has some suggestions. Drawing on ideas from information theory, he argues that more complex organisms would require physically longer descriptions than less complex organisms.

We have an intuitive sense that a lobster, say, is more complex (more "advanced", some might even say more "highly evolved") than another animal, perhaps a millipede. Can we *measure* something in order to confirm or deny our intuition? Without literally turning it into bits, we can make an approximate estimate of the information contents of the two bodies as follows. Imagine writing the book describing the lobster. Now write another book describing the millipede down to the same level of detail. Divide the word-count in the one book by the word-count in the other, and you will have an approximate estimate of the relative information content of lobster and millipede. It is important to specify that both books describe their respective animals "down to the same level of detail". Obviously, if we describe the millipede down to cellular detail, but stick to gross anatomical features in the case of the lobster, the millipede would come out ahead.

But if we do the test fairly, I'll bet the lobster book would come out longer than the millipede book. (Dawkins 2003, 100)

Frankly, I am not sure how easy it would be to do any of this. Suppose you compare a millipede with a warthog. What is the "same level of detail"? Do you count each leg of the millipede? If so, do you then count each hair of the warthog? Do you count the numbers of cells, or do you count the organization of cells? There was a time when it might have seemed plausible to count the number of genes, but now we know that organisms do more with less, in the sense that it is what you do with the functional genes that really counts, rather than the brute number. The genome size of some reptiles is far greater than that of humans. In any case, it is far from obvious to me that a human would come out ahead of an elephant or a whale, if all we are doing is simply describing them bit by bit.

A related problem of course is that even if you do find some measure of complexity, and even if it is one that puts humans at the top, it is not clear that this means that the complexity is better or more desirable, certainly not from a Darwinian perspective. McShea (1991) invites us to consider a tiger and a whale. By any measure the tiger's backbone is more complex than that of the whale. Moreover, although the whale did not evolve from a tiger, it did evolve from a land mammal with a backbone more in the tiger realm of complexity than its own realm. And yet no one – certainly no Darwinian – would say that the whale's backbone is adaptively inferior to that of the tiger. We all know that whales are superbly adapted creatures to their watery environment and the backbone is part of the story. Today, thanks to humans, whales are in trouble. But then so too are tigers. I don't think that people really have this in mind when they talk about adaptive excellence. Parenthetically, writing some twenty years later with Brandon, McShea does seem to accept some kind of complexity measure, and to argue that overall we see a rise in the notion; but, now he is pulling back from cherishing Darwinian criteria of excellence and is willing to go with a more neutral notion of complexity that seems basically just to involve lots of different pieces. There is no direct tie to humans and their intrinsic worth. Spencer, of course, did think that complexity leads up to humans and their great value, but how this happens exactly is rather left blank. (There is more on Spencer and progress in a later chapter.)

What about other notions of excellence perhaps not directly related to complexity? Paleontologist David Raup (1988) suggests that there might be some evidence that younger species in evolving lines live longer than older species. Again, though, this may all be a bit arbitrary. If you use other categories, like genera or families, then you run into anomalies. A living fossil like the horseshoe crab has been around for a very long time. Does one necessarily want to say that it is superior to a group that evolves more rapidly? Perhaps it has just found itself a very stable niche into which it fits nicely and there is no need of change. Other organisms live in environments that change more frequently and thus need to respond more quickly. Indeed, one might want to say that the organisms that do respond more are in some sense superior. This points in a fashion to another criterion that Dawkins proposes for excellence, namely the "evolution of evolvability." He argues that there are times when you just get evolutionary breakthroughs – like the eukaryotic cell – that have more potential. At such times, evolution makes a jump to a new dimension.

> Notwithstanding Gould's just skepticism over the tendency to label
> each era by its newest arrivals, there really is a good possibility that
> major innovations in embryological technique open up new vistas of
> evolutionary possibility and that these constitute genuinely progressive
> improvements (Dawkins 1989; Maynard Smith and Szathmáry 1995).
> The origin of the chromosome, of the bounded cell, of organized
> meiosis, diploidy and sex, of the eucaryotic cell, of multicellularity, of
> gastrulation, of molluscan torsion, of segmentation – each of these may
> have constituted a watershed event in the history of life. Not just in the

normal Darwinian sense of assisting individuals to survive and reproduce, but watershed in the sense of boosting evolution itself in ways that seem entitled to the label progressive. It may well be that after, say, the invention of multicellularity, or the invention of metamerism, evolution was never the same again. In this sense, there may be a one-way ratchet of progressive innovation in evolution. (Dawkins 1997, 1019–20)

Again the computer metaphor comes into play.

Computer evolution in human technology is enormously rapid and unmistakably progressive. It comes about through at least partly a kind of hardware/software coevolution. Advances in hardware are in step with advances in software. There is also software/software coevolution. Advances in software make possible not only improvements in short-term computational efficiency – although they certainly do that – they also make possible further advances in the evolution of the software. So the first point is just the sheer adaptedness of the advances of software make for efficient computing. The second point is the progressive thing. The advances of software open the door – again I wouldn't mind using the word "floodgates" in some instances – open the floodgates to further advances in software. (Ruse 1996, 469. This is from a presentation given in Melbu, Norway, in 1989.)

Explicitly Dawkins argues that evolution is cumulative. It has "the power to build new progress on the shoulders of earlier generations of progress." And brains, especially the biggest and best brains, are the epitome of adaptive excellence: "I was trying to suggest by my analogy with software/software coevolution, in brain evolution that these may have been advances that will come under the heading of the evolution of evolvability in [the] evolution of intelligence."

No doubt Jack Sepkoski would have had something to say about this, pointing out that brains are all very well, but they are very expensive to maintain. We prefer big brains, no doubt, but whether nature feels the same way is another matter. The AIDS virus is doing very well at the moment, whereas the chimpanzee in its natural habitat is deeply threatened. Is any Darwinian surprised? My suspicion grows that a Darwinian would be surprised only if someone did propose a notion of progressive excellence that worked! Certainly one can readily think of objections to other mooted proposals. For instance, Julian Huxley's demands for independence and control are surely weighted unfairly toward humans if the conclusion is that humans come top. Take the whale again. In major respects, the whale has independence and control that we never have. Only with huge amounts of equipment can we plumb the depths of the ocean, something that whales do with ease. Or take the camel. Despite the jokes about its being a horse designed by a committee, it is in fact superbly designed for desert life, able to survive for great lengths of time without water or other sustenance. Why, if we take into account habitats, should we judge it to be lower in the scale of being than humans?

None of this is to say that we humans should value fellow humans less than or even equal to other animals (or plants). I think all of us would think it morally wrong were someone to save a dog at the expense of a child. But it is to say that it is not readily apparent that Darwinian evolutionary theory sits comfortably with most proposed notions of biological progress. Perhaps an ecumenical conclusion is warranted. The eminent evolutionary biologist Francisco J. Ayala (1988) has said: "well, I would say that by many definitions, including very biologically meaningful definitions, humans are more progressive than any other organism," adding "'progressive' is an evaluative term that demands a subjective commitment to a particular standard of value" (95). Leave it at that.

As you might expect, with the very notion of progress in evolutionary biology being so messy, the various candidates for producing it are equally problematic – if not problematic as matters of pure science, then problematic as you try to unpack the philosophical ideas or intents behind them. There are some who are passionate defenders of the reality behind the metaphor of arms races. Invertebrate paleontologist Geerat J. Vermeij (1987) argues that we should and do see improvement over time. He writes: "if selection among individuals predominates over other processes of evolutionary change, and if enemies are the most important agencies of this selection, the incidence and expression of traits that enable individual organisms to cope with their enemies (competitors and predators) should be found to increase within specified habitats over the course of time." He adds: "On the whole, the evidence from fossils is in accord with the hypothesis of escalation" (359). Others, however, are not so sure. Vertebrate paleontologist Robert Bakker (1983) fails to find much evidence in the fossil record for the often cited example of prey getting faster and predators following suit. The trouble is that, even if there are arms races in nature, there is still a big gap between this and the matters that interest us here.

Vermeij lists such things as shell thickening, increased boring (of shells) ability, and so forth. Others have talked about the abilities of cuckoos to deceive their hosts and of the hosts to detect cuckoos and their eggs (Davies 1992). Although these may point to progress on some relative scale, none of them say much about progress on an absolute scale – an absolute scale with humans at the top. Darwin and Dawkins think that somehow superior intelligence will emerge. But this is surely a leap of faith (Ruse 2001). It is certainly not inevitable, because clearly the world might have been such that intelligence is simply too expensive to produce and maintain. More than this, one has a sneaking feeling that we are arguing too much from what has happened to what must happen. Without taking a hardline stand on the nature of body and mind, basically we seem to have the material and the thinking, the sentient. But this is our perspective. Who is to say that there is not a third option that might have emerged, perhaps along with sentience or as an alternative? That seems ridiculous to us, but that is the whole point. Of course it does! We build in a straight line to us. And that surely is what we should be proving, not assuming.

Similar worries seem to haunt the proposal of Simon Conway Morris. Critics (with some good reason) have objected to his whole scenario, arguing that his assumption that empty niches exist out there, objectively, is simply not well taken (Ruse 2008). Niches are a dynamic interaction between organisms and their environment. Take for instance the areas at the tops of trees in forests. Huge numbers of insects live up there, leading to a multiplicity of predators like birds. But in what sense can we say that such a niche was waiting there to be discovered, rather than being created by the growth of the trees in the crowded way that we find in forests and jungles? The point is not that there will be no niches, but that there is no guarantee that the niches we do have will exist or, at least as important from the viewpoint of progress, that they will exist in the kind of step-wise way that led to our own evolution. Conway Morris would of course reply that the existence of parallel niches, as for the saber-toothed tigers, shows that the niches are independent in some way. But is this really so? Marsupials and placental mammals have indeed evolved apart, but are they really that far apart – or are their environments that far apart? As with arms races, so much seems to be a matter of looking back and then finding that things that did happen were things that had to happen. Gould (1989) used to argue that if we think of life's history as a recorded tape (a metaphor which, I am afraid, dates him somewhat), we should not think that playing the tape again will always yield the same tunes. He may have had a point.

What then about complexity, which does seem to escape some of these worries? You are simply arguing that things will get more complex, perhaps in part thanks to selection, but also in large part because that is the way that things will go, physically, thanks to the nature of the world. Grant that this is so, although remember that everything is going to be dependent on a relative stability in planet Earth. Suppose, for instance, that everything does start to get hotter and hotter, or (as has happened) comets hit the Earth and stir up the atmosphere. It seems plausible to suppose that complexity might break down in at least some of these situations. Perhaps it will start building again, but if disruptions are ongoing, one would like a little more proof that this will happen. More worrisome is the old adage: what you gain on the roundabouts, you lose on the swings. If complexity increases, who is to say that it will be desirable, and, in particular, who is to say that it will lead to consciousness of the kind that we have? Short of a really good explanation of the body–mind relationship, which no one seems to have (and which some now doubt we could ever have), who is to say that there might not be some kind of complex functioning that does not involve sentience. What is more, if you put the main burden on complexity per se, and play down the significance of selection (which would seem to be Gould's gambit), then why should function – which seems to be a major reason for intelligence – play any significant role at all?

Let me sum up. There is no clear-cut answer to the question about the relationship between Darwinian evolutionary theory and hopes of progress, especially progress up to human beings. There are arguments suggesting that, despite the contingency of the Darwinian process, some kind of advance is not only possible but likely. I think few on the scientific side would argue for some sort of necessity. Perhaps, all other things being equal, complexity will inevitably increase over time. But are all other things always equal and, even if they are, does complexity in itself yield the kind of result (human-like qualities) that we want? Ultimately the only people who are really going to be upset by these conclusions are probably the religious, for whom the arrival of human beings is a no-compromise demand. This is not a scientific problem but a theological problem, and my suspicion is that believers should therefore search for some kind of theological solution. Supposedly scientific solutions, like that of Robert J.

Russell, that put God into the natural creative process generally (always?) turn out to be unsatisfactory for both science and theology.

For what it is worth, my own solution (at least for the Christian) is to play on the fact that God is outside time and space and that He can create as many universes as He wants (Ruse 2010). Since we have evolved, it was possible for us to evolve by natural causes. At some point, we were (or would be) bound to appear. God is not waiting around for this to happen, as would be the case for us. You might think that this is all an awful lot of hard work and waste – all of those non-productive universes. But really you have that already in our universe. All of those galaxies doing nothing productive, at least in the sense of producing human-like beings. In the nineteenth century there was a lot of angst about this apparent waste, leading those with a taste for natural theology like the Scottish physicist and biographer of Newton David Brewster and the historian and philosopher of science William Whewell to argue at length about possible solutions. One cannot truly say that much light was thrown on the issues, so perhaps these are theological worries we can leave to those who care. (See Whewell [2001], especially my introduction.)

In a way, those of us who take seriously David Hume's ([1739/40] 1940) point about the impossibility of deriving values from facts would have realized from the outside that the attempt to get progress out of the evolutionary process was in some sense doomed to failure. At most, one could get something that coincides with human nature but no guarantee that this something is of any worth. My own take is that we will probably never get quite what we want, but that no failure will quell the feeling that somehow there really is progress up to humankind (Ruse 1996). I think we are caught in a kind of Cartesian situation. If you doubt Descartes's *cogito ergo sum*, then you reaffirm its truth. To doubt is to think is to be. The same is true if you ask whether there is biological progress leading up to humans. On the one hand, you are drawing attention to the fact that we are still around and are among the last products of evolution. This in itself seems to give us a special place at the top of the tree. On the other hand, as Simpson pointed out, you are showing that you have the ability to ask about whether there is progress. This seems to give us special abilities. Either way, we won – and that is surely what progress is all about.

5   Knowledge

Charles Darwin was a scientist, not a philosopher. However, he was a well-educated Englishman from the first half of the nineteenth century and as such had a good grasp of the history of philosophy, something reinforced by family connections and friends, especially the London-based set in which his elder brother (another Erasmus Darwin) moved. He had things to say both about epistemology (the theory of knowledge) and ethics (the theory of morality). The former topic is the subject of this chapter and the latter of the next chapter. In both cases we begin with Darwin, but the aim is to move to the present and to ask about the importance of Darwinian evolutionary theory (as conceived today) to contemporary philosophical thinking. I should say that were I writing this book some seventy or eighty years ago, the treatment would be brief and critical. The highly influential, Austrian-born, England-residing philosopher Ludwig Wittgenstein spoke for many when he said, somewhat contemptuously: "Darwin's theory has no more relevance for philosophy than any other theory in natural science" (Wittgenstein 1923, 4.1122). General opinion was that, if anything, bringing Darwin to bear on these issues was not just wrong, but somehow intellectually unclean. A bit like a bad smell at a vicarage garden party. People who got into this business probably had some unspoken agenda, either of a mystical or more dangerous political nature. One confesses that there was some justification for this attitude. But while it is important to understand the past, we do not necessarily have to be its prisoners. Today there is a huge amount of interest in the relevance of evolutionary thinking to philosophical issues, and even the skeptics agree that the interaction is beneficial. Let us see why. (A recent reader, Ruse 2009b, contains much material pertinent to this and the next chapter.)

The evolution of ideas

Darwin had much more to say about ethics than about epistemology, but what he did have to say about the latter contains the germs of subsequent thinking. There seem to be two major approaches to the application of Darwinian thinking to epistemology. The first is more analogical. The second is more literal. Starting with the first, we turn to the *Descent of Man*, which as we know appeared in 1871, twelve years after the *Origin*. It is worth quoting Darwin in full.

> The formation of different languages and of distinct species, and the proofs that both have been developed through a gradual process, are curiously the same. But we can trace the origin of many words further back than in the case of species, for we can perceive that they have arisen from the imitation of various sounds, as in alliterative poetry. We find in distinct languages striking homologies due to community of descent, and analogies due to a similar process of formation. The manner in which certain letters or sounds change when others change is very like correlated growth. We have in both cases the reduplication of parts, the effects of long-continued use, and so forth. The frequent presence of rudiments, both in languages and in species, is still more remarkable. The letter *m* in the word *am*, means *I*; so that in the expression *I am*, a superfluous and useless rudiment has been retained. In the spelling also of words, letters often remain as the rudiments of ancient forms of pronunciation. Languages, like organic beings, can be classed in groups under groups; and they can be classed either naturally according to descent, or artificially by other characters. Dominant languages and dialects spread widely and lead to the gradual extinction of other tongues. A language, like a species, when once extinct, never, as Sir C. Lyell remarks, reappears. The same language never has two birth-places. Distinct languages may be crossed or blended together. We see variability in every tongue, and new words are continually cropping up; but as there is a limit to the powers of the memory, single words, like whole languages, gradually become extinct. As Max Müller has well remarked: – "A struggle for life is constantly going on amongst the words and grammatical forms in each language. The better, the shorter, the easier forms are constantly gaining the upper hand, and they owe their success to their own inherent virtue." To these more important causes of the survival of certain words, mere novelty may, I think, be added; for

> there is in the mind of man a strong love for slight changes in all things. The survival or preservation of certain favoured words in the struggle for existence is natural selection. (C. Darwin 1871, I, 60)

This may indeed be natural selection, but notice that it is not the literal success of one organism over another, but rather of one word over another. And of course, words do not truly triumph as such, but in the use of one speaker over another, or the same speaker at different times. In other words, we have an analogy or metaphor. For this reason, note that we have gone beyond the cultural evolutionists discussed in chapter 3. Now the biological is right out of the picture. Note also that Darwin is talking here about language. This was picked up by others. One such was the early American pragmatist, Chauncey Wright – a fanatical Darwinian and much appreciated by the master himself, who had reprinted at his own expense one of Wright's reviews of a book very critical of evolution by natural selection. He wrote, "in the development of language, its separations into the varieties of dialects, the divergences of these into species, or distinct languages, and the affinities of them as grouped by the glossologist into genera of languages, present precise parallels to the developments and relations in the organic world which the theory of natural selection supposes" (C. Wright 1877, 257). However, there is no need to restrict the analogy to language alone. The better-known pragmatist, William James, used the analogy to talk about influential individuals in a society.

> The causes of production of great men lie in a sphere wholly inaccessible to the social philosopher. He must simply accept geniuses as data, just as Darwin accepts his spontaneous variations. For him, as for Darwin, the only problem is, these data being given, how does the environment affect them, and how do they affect the environment? Now, I affirm that the relation of the visible environment to the great man is in the main exactly what it is to the "variation" in the Darwinian philosophy. It chiefly adopts or rejects, preserves or destroys, in short *selects* him. And whenever it adopts and preserves the great man, it becomes modified by his influence in an entirely original and peculiar way. He acts as a ferment, and changes its constitution, just as the advent of a new zoölogical species changes the faunal and floral equilibrium of the region in which it appears. (James 1880b, 445)

And:

> The mutations of societies, then, from generation to generation, are in the main due directly or indirectly to the acts or the examples of individuals whose genius was so adapted to the receptivities of the moment, or whose accidental position of authority was so critical that they became ferments, initiators of movements, setters of precedent or fashion, centers of corruption, or destroyers of other persons, whose gifts, had they had free play, would have led society in another direction. (445)

More broadly, James saw the Darwinian process going on in the very fabric of thinking, for "the new conceptions, emotions, and active tendencies which evolve are originally produced in the shape of random images, fancies, accidental out-births of spontaneous variation in the functional activity of the excessively instable human brain, which the outer environment simply confirms or refutes, adopts or rejects, preserves or destroys, – selects, in short, just as it selects morphological and social variations due to molecular accidents of an analogous sort" (p. 456). Just as there is a struggle for existence, with consequent evolution in the direction of adaptive excellence in the world of animals and plants, so likewise there is a struggle for existence between ideas, with consequent evolution in the direction of adaptive excellence in the world of knowledge.

Notice why the pragmatists might be expected to turn with enthusiasm to Darwinism. There is a kind of relativism about both systems. What succeeds is what succeeds. It was this as much as anything that led to the decline of such analogizing in the first part of the twentieth century. Bertrand Russell, one of the greatest influences on the development of "analytic" philosophy – the dominant approach in anglophone societies – saw pragmatism as epistemologically fallacious and moral and socially dangerous.

> Pragmatism, in some of its forms, is a power-philosophy. For pragmatism, a belief is "true" if its consequences are pleasant. Now human beings can make the consequences of a belief pleasant or unpleasant. Belief in the moral superiority of a dictator has pleasanter consequences than disbelief, if you live under his government. Wherever there is effective persecution, the official creed is "true" in the pragmatist sense. The pragmatist philosophy, therefore, gives to those in power a metaphysical omnipotence which a more pedestrian philosophy would deny to them. (Russell [1937] 2004, 174)

Pragmatism was out of fashion for many years. However, around 1960, things started to change, no doubt in part because by then Darwinism was an up and running theory of science, and natural selection was seen by professional scientists as the key mechanism in evolutionary change. The analogy between the struggle among organisms and the struggle among ideas started to flourish again, particularly in the hands of those who were interested in the development of science itself. Karl Popper (1974) was one who floated the connection, although perhaps typically he did so more to legitimate his own supposedly independently discovered insights than to promote a version of Darwinism itself. As is well known, Popper argued that scientists (that is to say, good scientists doing what good scientists should do) throw up hypotheses or conjectures and then they try to refute them, to falsify them. If they cannot, then the hypotheses last for a while. If they can, then new hypotheses are suggested and the scientific process starts all over again. Schematically:

$P_1 \rightarrow TS \rightarrow RR \rightarrow P_2$
(P_1 is an existing problem, TS is a hypothesis or tentative solution, RR is a successful attempt at falsification or rigorous refutation, and P_2 is a new problem.)

Somewhat ironically, Popper (correctly) saw that what he was offering was not a scientific theory but a philosophical or metaphysical proposal. Popper (incorrectly) also saw the analogy between what he was offering and what Darwinism offers as strengthened by the fact that the latter is no genuine scientific theory but a "metaphysical research programme." Perhaps overly influenced by the social uses to which Darwinism was put in the Austria of his youth (more on this in the next chapter), Popper was never truly comfortable with Darwinism as science, and to the end yearned for something a bit more teleological – doubly ironical for a man who criticized Hegelian, directed theories of history (Beatty 2001).

A more comfortable use of the analogy came from the pen of the British-born philosopher Stephen Toulmin.

Science develops ... as the outcome of a double process: at each stage a pool of competing intellectual variants is in circulation, and in each generation a selection process is going on, by which certain of these variants are accepted and incorporated into the science concerned, to be passed on to the next generation of workers as integral elements of the tradition.

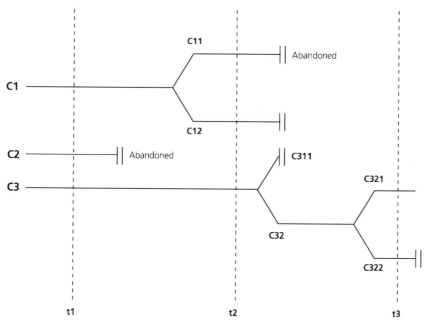

Figure 5.1 The evolution of theories as envisioned by Stephen Toulmin.

Looked at in these terms, a particular scientific discipline – say, atomic physics – needs to be thought of, not as the contents of a textbook bearing any specific date, but rather as a developing subject having a continuing identity through time, and characterized as much by its process of growth as by the content of any one historical cross-section … Moving from one historical cross-section to the next, the actual ideas transmitted display neither a complete breach at any point – the idea of absolute "scientific revolutions" involves an over-simplification – nor perfect replication, either. The change from one cross-section to the next is an *evolutionary* one in this sense too: that later intellectual cross-sections of a tradition reproduce the content of their immediate predecessors, as modified by those particular of the professional standards of the science of the time. (1967, 465–66)

And he offered a detailed picture to illustrate his thinking.

Problems

Whatever the faults of this approach to what has been called "evolutionary epistemology" – and I shall be looking at some critics in a moment – what cannot be denied is that it has inspired some innovative history of science.

David Hull (1988) has written a stimulating analysis of the battles in the 1970s that occurred in the field of systematics when the new taxonomic practice of "cladism" was introduced, and Robert J. Richards (1987) wrote a no-less important account of the significance of Darwinian thinking on the development of theories of the mind. Yet this in itself is obviously not enough. One might still love *War and Peace* and yet reject Tolstoy's critique of the great-man theory of history. That the analogy might not be entirely satisfactory is shown perhaps by the ease with which people of very different philosophical persuasions feel able comfortably to slip beneath its mantle. Popper's great philosophical opponent was Thomas Kuhn (1962). Instead of seeing science as a never-ending struggle through conjecture and refutation to a partial understanding of objective reality, he argued that science makes reality and that we are always caught in frameworks, paradigms, that cannot be challenged by reason and evidence and can only be changed by conversion experiences akin to a religious or political road-to-Damascus experience. Yet, no less than Popper, he claimed to be a Darwinian! You might think that this is a classic case of trying to light the fire after you have soaked the altar with water. After all, what could be further from Darwinian smooth continuity than what happened to St. Paul when he was struck blind, or what supposedly happens to a scientist when he or she swops paradigms? Kuhn, however, was not in the business of trying to square the circle, for he concentrated on a different aspect of the Darwinian picture – he saw Darwinism as challenging fundamental ideas and hopes of absolute progress. Since he thought Darwinism allowed but a relative progress, that is what he wanted for science. Paradigm switches are not and cannot be moves closer to objective reality, because a paradigm defines reality itself. He therefore wrote:

> It helps to recognize that the conceptual transposition here recommended is very close to the one that the West undertook just a century ago.
> It is particularly helpful because in both cases the main obstacle to transposition is the same ... The "idea" of man and of the contemporary flora and fauna was thought to have been present from the first creation of life, perhaps in the mind of God. That idea or plan had provided the direction and the guiding force to the entire evolutionary process. Each new stage of evolutionary development was a more perfect realization of a plan that had been present from the start.
>
> For many men the abolition of that teleological kind of evolution was the most significant and least palatable of Darwin's suggestions. The

> *Origin of Species* recognized no goal set by God or nature. Instead, natural
> selection, operating in the given environment and operating with the
> actual organisms presently at hand, was responsible for the gradual but
> steady emergence of more elaborate, further articulated, and vastly more
> specialized organisms. Even such marvelously adapted organs as the eye
> and hand of man – organs whose design had previously provided powerful
> arguments for the existence of a supreme artificer and advance plan –
> were products of a process that moved steadily from primitive beginnings,
> but toward no goal. (Kuhn 1962, 172)

Whether or not we want to accept Kuhn's own picture of scientific change, we can agree that he does have a point. We know full well that clear-cut notions of absolute biological progress are problematic, to say the least. And yet most people would want to agree with Popper that science is about an objective reality – a knowledge without a knower. We may never get there, but progress in science is absolute in some sense. Related to this is the fact that, for all of the genetic exchange that might occur across very different branches of life, by and large Darwinian evolution is a matter of branches getting farther and farther apart. Humans are not mushrooms. But in science, the major triumphs come when different branches are united under one hypothesis. That is what the consilience of inductions is all about, and that, as Whewell rightly noted, is just about the best evidence that we have that we are dealing with a non-human-based objectivity.

Ultimately, the trouble seems to be in the nature of variation (Ruse 1986). The essence of Darwinism is that the new variations, the mutations in today's language, are random, in the sense that they do not occur according to need. Even if they are just what is required, there was no design behind their appearance. And yet although it is certainly possible that the new variations of science be random, the story of Alexander Fleming and the discovery of penicillin comes to mind, by and large they are the epitome of design. They are sought after and come to the best, prepared minds. Charles Darwin and his eighteen-month struggle to find natural selection is an obvious case. The great pragmatist Charles Sanders Peirce saw the problem.

> As for the … mode of development which I have called Darwinism,
> however important it may be in reference to some of the growths of
> mind – and I will say that in my opinion we should find it a considerable
> factor in individual thinking – yet in the history of science it has been

made as far as we have been able to see, no figure at all, except in retrograde movements. In all these cases it betrays itself infallibly by its two symptoms of proceeding by insensible steps and of proceeding in a direction different from that of any strivings. (Peirce [1892] 1958, 257)

What to say in response? The least satisfactory reply seems to me that offered by Popper. He tried to find some naturalistic way of putting direction into the new variations, via a kind of Lamarckian inheritance of acquired behaviors. Even those sympathetic to his philosophy saw that this would not do, and soon he himself backed down from this strategy. More satisfactory perhaps is to take the line adopted by David Hull, namely to agree that there are fundamental differences between the variations of life and those of science, but to point out (reasonably) that all analogies have points of dissimilarity – otherwise they would not be analogies! This is true, although of course the downside is that the greater the dissimilarity, the less useful the analogy appears. It is true that Hull himself had great success using the analogy to write his history of modern taxonomy, although an unkind critic might question whether he had chosen as a topic a somewhat atypical branch of science. The debate among taxonomists was not so much about objective reality, but how to classify it. In other words, we were already in the realm of subjective ideas, and this lent itself readily to a philosophy where subjective ideas were paramount. Perhaps best of all is to think of this kind of evolutionary epistemology less as a full-blown theory of evolutionary change, and more as a proposal that will have to wait for full success on the development of an adequate theory of cultural evolution. For the time being, we can take it as an analogy or metaphor that leads to interesting insights and a stimulus to further thought. For instance, if like Kuhn you are worried about claims supposing a human-independent objective reality – and we shall be raising this assumption again soon – then you might find illuminating the analogy between Darwinian change and scientific change (that is, if you think that absolute biological progress is impossible). But whether you want to say that they are essentially the same process seems to me to be another matter.

Darwin on the mind and knowledge

Let us change tack now, and look at the other approach that uses Darwinism to solve issues about the theory of knowledge. We have looked at the analogical route. Now let us look at the literal route. And to do so, let us return to Darwin and this time to the notebooks that he kept in the late 1830s, while discovering and working out his evolutionary thinking. These are the important source. As noted earlier in this book, one of the most interesting things about Darwin as an evolutionist is his stone-cold certainty that any theory must apply completely and utterly to our own species. I have suggested that this probably has roots in his experience on the *Beagle*, when he saw the Fuegians in their natural habitat, and more importantly how the seemingly civilized trio that were returning from Britain almost at once reverted to type. Even the most sophisticated of humans are but a hairsbreadth from the savage, and from there it is only a step to the brute. Interestingly and suggestively, the very first intimation that we have in the notebooks that Darwin really has grasped natural selection as a mechanism comes a month or so after the crucial reading of Malthus, and it is about humankind. "An habitual action must some way affect the brain in a manner which can be transmitted.–this is analogous to a blacksmith having children with strong arms.– The other principle of those children. which *chance?* produced with strong arms, outliving the weaker one, may be applicable to the formation of instincts, independently of habits.–" (Notebook entry, November 27, 1838, N 42, in Barrett *et al.* 1987, 574).

Darwin does not have a particular theory of the relationship between body and mind, but he is clear that the mind is in some sense part of the material world, and as such subject to natural selection. More particularly, as an evolutionist he denies that the mind is a *tabula rasa*, that is to say a completely blank sheet on which experience writes the text. It is obvious to Darwin that we come into the world with (to use a modern metaphor) a huge amount of software already installed – software that was of importance to our ancestors in the struggle for existence. "Origin of man now proved. – Metaphysic must flourish. – he who understands baboon would do more towards metaphysics than Locke" (M 84e, August 16, 1838, Barrett *et al.* 1987, 539, that is before the discovery of selection). Then a few weeks later, he lets us see explicitly the line that he is following. Instinct formed by evolution gives us innate knowledge. "Plato says in Phaedo that our '*necessary ideas*' arise from the preexistence of the soul, are not derivable from experience. – read monkeys for preexistence – " (M 128, September 4, 1838, Barrett *et al.* 1987, 551). In this notebook, Darwin makes it clear that he is relying here on his elder brother for the information about Plato.

Showing that he was a scientist and not a philosopher, Darwin did not follow through these speculations. But he did return once more to the topic, a couple of years later in 1840. He was reading an article by the philosopher John Stuart Mill in one of the most important contemporary journals, the *Westminster Review*. Mill (1840) was explaining the difference between the empiricists like the English philosopher John Locke, who think that all knowledge is based on experience, and those like the German philosopher Immanuel Kant, who think that the mind structures experience. The poet Samuel Coleridge was instrumental in bringing Kant's thinking to the attention of the British intellectual public.

> Every consistent scheme of philosophy requires, as its starting point, a theory respecting the sources of human knowledge ... The prevailing theory in the eighteenth century was that proclaimed by Locke, and attributed to Aristotle – that all our knowledge consists of generalizations from experience ... From this doctrine Coleridge with ... Kant ... strongly dissents ... He distinguishes in the human intellect two faculties ... Understanding and Reason. The former faculty judges of phenomena, or the appearance of things, and forms generalizations from these: to the latter it belongs, by direct intuition, to perceive things, and recognize truths, not cognizable by our senses. (Mill 1840, 267)

Mill made it very clear that he himself was in the empiricist camp. "We see no ground for believing that anything can be the object of our knowledge except our experience, and what can be inferred from our experience by the analogies of experience itself." (Mill 1840, 267) Darwin took this as an opportunity to reassert his evolutionary position. He jotted down his thinking on a page of a bundle that today is somewhat misleadingly known as "Old and Useless Notes." (Misleading because this is a later title, and given by Darwin in old age when he was searching through earlier writings for material still to be utilized.)

> Westminster Review, March 1840 p.267 – says the great division amongst metaphysicians – the school of Locke, Bentham, & Hartley, &. and the school of Kant. to Coleridge, is regarding the sources of knowledge. – whether "anything can be the object of our knowledge except our

experience". – is this not almost a question whether we have any instincts, or rather the amount of our instincts – surely in animals according to the usual definition, there is much knowledge without experience. so there *may* be in men – which the reviewer seems to doubt. (OUN 33, in Barrett *et al.* 1987, 610)

The basic idea is simple. We are born with innate knowledge that was forged in the struggle for existence. Those of our would-be ancestors who had such knowledge are now our real ancestors. Those who did not are not. Nobody of course thinks that this knowledge comes automatically. John Locke, the great opponent of innate knowledge, noted that no one is born knowing that God exists or that parents should love their children. The point is, rather, that in the course of development in culture, certain dispositions channel our understanding one way rather than another. As far as I know, although he was obviously committed to the ideas, Darwin never published on this topic. I don't think there is any great significance to this. He had lots of other fish to fry. Including morality where, as we shall see, he offered a full discussion. In any case, the topic was being taken up with great gusto by another, namely Herbert Spencer. In his *Principles of Psychology*, published some four years before the *Origin*, with a certain justice Spencer thought he was steering a mid-course between the empiricists, who insisted on the primacy of experience, and Kant, who argued for the mind-informing element in knowledge. Like Darwin, Spencer also saw the mind as structured by innate principles or dispositions. Like Kant, Spencer identified these principles with basic mathematical claims and with the underlying principles of scientific thinking, such as the need to see all phenomena and events as having causes. Where he differed from Darwin was in seeing the force behind this picture of the structured mind as essentially a question of Lamarckian inheritance – even into the twentieth century, Spencer's favored mechanism of change. Where he went beyond Darwin – although it may well have been that Darwin would have been happy to follow – was in tying this all to a firmly Progressionist picture of knowledge. As in all things, as scientists work away at their tasks, we gain an ever-more accurate understanding of absolute reality. With an atypical burst of modesty and a nod to the Kantian thing-in-itself, Spencer did, however, refer to the underlying reality as the (capitalized) Unknowable or Absolute.

In Darwin's footsteps

Again such ideas lay fallow and were certainly not to be favored by right-thinking philosophers in the early twentieth century. One who did seize on them was the ethologist Konrad Lorenz (1941), typically a biologist and not a philosopher. He gave a highly Kantian version of this approach, arguing that now we can see what Kant had regarded as in some sense necessary, synthetic a priori conditions of thinking, are in fact innate principles put in place by natural selection. Like Kant, Lorenz was an out-and-out realist, thinking that the physical world exists independently of humans and their perceptions. "Our categories and forms of perception, fixed prior to individual experience, are adapted to the external world for exactly the same reasons as the hoof of the horse is already adapted to the ground of the steppe before the horse is born and the fin of the fish is adapted to the water before the fish hatches" (233). Where Lorenz thought that he was bringing Kant up to date via evolutionary biology was in claiming that we can (contrary to what Kant thought) get some real understanding of this human-independent world – the "thing-in-itself," or the *Ding an sich.* "Our categories and forms of perception, fixed prior to individual experience, are adapted to the external world for exactly the same reasons as the hoof of the horse is already adapted to the ground of the steppe before the horse is born and the fin of the fish is adapted to the water before the fish hatches" (233). However, where Kant went really wrong was in thinking that these categories and forms are something rather metaphysical, necessary for any thought at all. As noted, they are, however, empirical constraints put in place by evolution. Lorenz writes of them as "inherited working hypotheses," which have shown their mettle in our dealing with the physical world. "This conception, it is true, destroys our faith in the absolute truth of any a priori thesis necessary for thought. On the other hand it gives the conviction that something actual 'adequately corresponds' to every phenomenon in our world" (239). Note that this means that we humans may not have achieved the highest form of knowledge. There may be beings standing in relationship to us as we stand in relationship to the water shrew. "To declare man absolute, to assert that any imaginable rational being, even angels, would have to be limited to the laws of thought of Homo sapiens, appears to us to be incomprehensible arrogance" (246).

Lorenz wrote in 1940, from the German side, in the early years of the Second World War. It is no surprise that it was many years before his thinking became widely known in the West. (This occurred around 1970.) By that time, others had started to move in the same direction. One was the doyen of mid-twentieth-century American philosophers, W. V. O. Quine (1969), who may well have been showing a certain sympathy for the legacy of the pragmatists, although perhaps even more that of David Hume. He wrote about the venerable problem of induction: why should the present and the future be like the past? We expect the sun to rise tomorrow because it always has done in the past, but logically this is not necessary. As Bertrand Russell used to say, perhaps we are like the turkey that has been fed every morning by the farmer, and now Christmas Eve has arrived and we are anxiously expecting breakfast! Hume's solution was to let psychology step in where philosophy had led to skepticism.

> Most fortunately it happens, that since reason is incapable of dispelling these clouds, nature herself suffices to that purpose, and cures me of this philosophical melancholy and delirium, either by relaxing this bent of mind, or by some avocation, and lively impression of my senses, which obliterate all these chimeras. I dine, I play a game of backgammon, I converse, and am merry with my friends; and when after three or four hours' amusement, I would return to these speculations, they appear so cold, and strained, and ridiculous, that I cannot find in my heart to enter into them any farther. (Hume [1739/40] 1940, 175).

Quine agreed that there may be no logical explanation for the success of induction, and taking things one step beyond Hume (who I do not think was an evolutionist but I think would have been sympathetic to the idea) argued that Darwinian theory gives the reason why our psychology functions as it does.

> One part of the problem of induction, that part that asks why there should be regularities in nature at all, can, I think, be dismissed. *That* there are or have been regularities, for whatever reason, is an established fact of science; and we cannot ask better than that. *Why* there have been regularities is an obscure question, for it is hard to see what would count as an answer. What does make clear sense is this other part of the problem of induction: why does our innate subjective spacing of qualities accord so well with the functionally relevant groupings in nature as to

make our inductions come out right? Why should our subjective spacing of qualities have a special purchase on nature and a lien on the future?

There is some encouragement in Darwin. If people's innate spacing of qualities is a gene-linked trait, then the spacing that has made for the most successful inductions will have tended to predominate through natural selection. Creatures inveterately wrong in their inductions have a pathetic but praise-worthy tendency to die before reproducing their kind. (Quine 1969, 126)

Working in the same vein, the philosopher Elliott Sober (1981), Quine's student, argued against the claim that "an evolutionary account of the origins of rationality is impossible because natural selection is too coarse-grained a process to single out the scientific method from innumerable other, less rational, procedures for constructing beliefs out of other beliefs" (95). Sober went beyond the general problem of induction to focus on some of the paradoxes of induction and confirmation that so fascinated philosophers in the 1960s. We are invited to consider the predicates "green" and "grue," where the latter is defined as "green before a time t and blue after," and t is some point in the future. It had been pointed out by people like Nelson Goodman (1955) that all of the green emeralds that we have seen thus far in our lives support both "all emeralds are green" and "all emeralds are grue." But surely we want to deny the latter? Invoking the notion of simplicity – at some level the predicate "green" is simpler than the predicate "grue" – Sober pointed out that this feeds nicely into a selection solution – formally there may be no difference in the inferences but it is a lot simpler, needs a lot less computing power, to go green rather than grue. It is not a question of absolute right and wrong but more one of pragmatics. The proto-human who sat around worrying about grue may have been the better philosopher but it is doubtful that he or she was fitter in a Darwinian sense than the simple-minded proto who was satisfied with thinking green.

This is all very much an empirical, or as it is generally called today a "naturalistic," approach to philosophy. One is setting out explicitly to make the findings and truths of science an integral part of the solution to the nature of knowledge. This then means, obviously, that you need to take the science seriously. If you are the kind of philosopher (a bit like Popper) who tells scientists what they should do – one is prescribing good scientific practice – then the fact that scientists do not in fact do what one says is not fatal to one's philosophy. They ought to do it, even if they don't! But if you are a naturalist, then you are more into description. You must start with what the scientist says. This point has bite when you start to think about what the precise nature of the innate dispositions that you think evolution has put in place to shape human thinking would be. In the specific case of science, I suppose one thinks most obviously of the basic principles of logic – the law of excluded middle, for instance – and of mathematics – the basic principles of arithmetic, for instance. This all seems to make good sense. The proto-human who refused to enter a cave because he saw three tigers go in and only two come out is going to be ahead of the proto who happily entered the cave for a good night's sleep. Also, staying with science, one thinks of what in the trade are called "epistemic principles" or "values." These are the sorts of criteria that philosophers have identified as characteristic of good science. A classic case of an epistemic value is the Whewellian consilience. Good science unifies different branches of understanding and lets you make inferences from hypotheses based on the evidence, the clues, of the different branches. Again, this seems to make good sense from a Darwinian perspective. The proto-human who went down to the water hole and found bloodstains and broken grass and heard growls in the undergrowth and who then said "Tigers, beware," is likely to be ahead of the proto who simply said "Tigers, just a theory not a fact."

Of course, if biology is any guide, one expects things to be a bit more complex than this and perhaps not as neat and tidy as philosophers might expect and hope. After all, if all of this is innate, why then do we have to spend so much time teaching our young the principles of reasoning and of mathematics? Those working on these sorts of issues stress that first and foremost we must remember that humans are a social species and that what would really be important would be reasoning abilities for social situations rather than reasoning abilities for passing logic exams. Leda Cosmides and John Tooby (1992, 1996, 2005) have been pioneers in stressing this point, and Cosmides (1989) in particular has tried to give empirical force to her concerns. She has focused her attention on those puzzles that so delight psychologists. Particularly pertinent are those paradoxes where humans perform well on one task and badly on another task, even though formally they are identical. The test devised by psychologist Peter Wason is a good example. Take a set of cards with letters on one side and numbers on the other. You are shown four cards with the distribution D; F; 3; 7. You are now asked the question, "Which cards must you turn over to see if the following rule holds true: 'If a card has a D on the one side, it must have a 3 on the other'?" Now try this one: "Given a rule that a bar bans under-18-year-olds drinking, to avoid having anyone break the rule, which of four customers must you check: the beer-drinker, the lemonade-drinker, the 25-year-old, or the 16-year-old?" Everyone gets the second problem right but most people flunk the first – even though, formally, the two situations are exactly the same. (In the first problem, most people chose D, which is correct, and 3, which is incorrect. They should have chosen 7. The rule doesn't say that only D leads to 3, but the rule does say that 7 cannot correspond to D. Everyone checks the beer-drinker's age and the 16-year-old's drink.)

Why the difference in performance? Simply because, in everyday life, we much more commonly encounter the boozing-type situation than the abstract number–letter situation, and so are better at solving it. This leads Cosmides to argue that the brain is not a simple all-purpose computer but one that reflects the needs of its possessors as they strive for success in life's struggles. Through a great deal of trial and error and sheer hard slog by scientists and others, we have worked out the basic principles of correct reasoning, and they are obviously reflected at some level in our innate dispositions. But there is certainly no simple one-to-one correspondence.

Other cautions or complexities need to be added. First, even if basic principles are in some sense embedded in the mind, no one thinks that the more complex derivations are very close to our biology. Take something like the Euler identity: $e^{\pi i} + 1 = 0$. I doubt very much that any proto-human ever had need of this, and if indeed it had been discovered back then (truly a ridiculous supposition) one suspects that thinking about it might have been a sign that one was truly out of touch and a Darwinian disaster. Science, that is to say modern science and technology, is clearly something that builds on and in major respects transcends its roots. Obviously – a point made in an earlier chapter about culture generally – in some broad sense the Darwinian success of humans is a function of science and technology (including medicine). Notwithstanding the Darwinian dangers to us all if our knowledge of nuclear forces leads to an atomic conflagration consuming the Earth, at a more immediate level we would not be around in the numbers that we are without science and technology. Think of the

Industrial Revolution and the consequent growth in population, or how in our own time the green revolution has saved so many lives. But it would be naïve to think that particular elements of a complex scientific theory are going to have direct Darwinian connections and benefits. Charles Darwin had ten children, seven of whom grew to maturity. Gregor Mendel, a priest, had none. But it would be silly to think that this is therefore a refutation of Darwinian theory because Mendel got genetics right and Darwin got it wrong.

Metaphor

Connected with this is a second point, namely that no scientific theory is just rules and valued methods of reasoning. There is not only empirical content, but also the ways in which one tries to conceptualize the empirical content. To do this, as many have noted, scientists turn to models and more particularly to metaphors. Think about Darwin's theory and its crucial dependence on metaphor – struggle for existence, natural selection, tree of life, and more recently adaptive landscape, genetic code, arms race. These are not just raw descriptions of reality, but ways of taking it up and conceptualizing it and giving it meaning and thus leading to understanding. Organisms are born, live, reproduce, and die. The metaphor of natural selection tells you what that is all about and why it matters. As we saw in chapter 3, sometimes in the name of some kind of ideological purity there are calls for the elimination of metaphor, but that is probably impossible and just stupid anyway. Apart from anything else, metaphors have incredible heuristic powers, and that is one of the most prized epistemic virtues. A theory that does not point you in new directions is a theory not worth having. Taking as an example a favored metaphor of Darwin's, the division of labor, we saw in chapter 1 that even to this day this plays a vital role in evolutionary theorizing. Edward O. Wilson has used it most profitably to look in some detail at the caste structure in the Brazilian leaf-cutter ants, the *Atti* (E. Wilson 1980a, 1980b). He shows in detail how the ants have divided up the work – soldiers and foragers and leaf-cutters and (back in the nest where they grow a fungus to feed to their young on the chewed-up leaves) gardeners, and gatherers, and nurses, not to mention the queen – exemplifying a highly sophisticated division of labor. The concept of the arms race, another metaphor already encountered, has likewise led to really important discoveries. Ornithologist Nicholas Davies (1992) has studied in some detail the adaptations that the cuckoo has developed to lay its eggs in the nests of other species and the corresponding adaptations developed by potential hosts to avoid such parasitism. Most obviously there are the egg colors. In Britain, different strands (gentes) of the same cuckoo species parasitize different birds – meadow pipits, reed warblers, dunnocks (hedge sparrows), robins (the British robin, not the bigger American one), and pied wagtails. Hosts are incredibly sensitized to egg color and will reject at once any egg that does not match exactly. Score one for the hosts. The eggs of the different gentes vary precisely to match those of the hosts. Score one for the cuckoos. Hosts reject eggs that are laid before they produce their own eggs. Score two for the hosts. Cuckoos are careful to lay eggs only when there are eggs. Score two for the cuckoos. Hosts get very tense when potential threats are around. Score three for the hosts. Cuckoos lay their eggs in fractions of the time most birds take to lay eggs. Score three for the cuckoos. And so it goes on. (Actually, dunnocks are an exception and show few adaptations to protect themselves from parasitism. They are, however, the exception that proves the rule because they are believed to have been parasitized only recently and a full arms race has not had time to kick in.)

Metaphors are crucial for science. To use a metaphor about metaphors, if we think of the epistemic rules and the principles of logic and mathematics as the skeleton of the body of science, then the metaphors add the flesh and skin. But notice that not any old metaphors will do. Yes, Darwin does use metaphors like the struggle for existence and natural selection, but they work and are cherished only insofar as they fit with and obey and promote the knowledge-producing values of science. If natural selection and the tree of life and so forth had not led to the consilience and had not shown that it works very fruitfully, no one would have taken Darwin's theory seriously, and rightly so. So what we have clearly is an intricate melding of biology and culture. Scientific theories are based on biology. Scientific theories are part of culture.

Naturalism self-refuting?

We cannot simply end on this cheerful note, for as with the other form of evolutionary epistemology, there are critics of this kind of approach also.

The concern is basically about why we should take seriously the products of evolution as guides to the ultimate nature of things. Although as part of his dislike of things English there was no way that the nineteenth-century German philosopher Friedrich Nietzsche was going to acknowledge a debt to that crude thinker Charles Darwin, nevertheless he adopted an evolutionary approach to philosophy not unlike that of the father of natural selection.

> It is high time to replace the Kantian question, "How are [a priori moral judgments] possible?" by another question, "Why is belief in such judgments *necessary*?" – and to comprehend that such judgments must be *believed* to be true, for the sake of the preservation of creatures like ourselves; though they might of course be *false* judgments for all that! ... After having looked long enough between the philosopher's lines and fingers, I say to myself: by far the greater part of conscious thinking must be included among instinctive activities, and that goes even for philosophical thinking. We have to relearn here, as one has had to learn about heredity and what is "innate" ... "Being conscious" is not in any decisive sense the opposite of what is instinctive: most of the conscious thinking of a philosopher is secretly guided and forced into certain channels by his instincts. (Nietzsche 1886, I, ii, 3)

Note how Nietzsche picked up on the possible falsity of such beliefs. He elaborated on this:

> Over immense periods of time the intellect produced nothing but errors. A few of these proved to be useful and helped to preserve the species: those who hit upon or inherited these had better luck in their struggle for themselves and their progeny. Such erroneous articles of faith, which were continuously inherited, until they became almost part of the basic endowment of the species, include the following: that there are enduring things; that there are equal things; that there are things, substances, bodies; that a thing is what it appears to be; that our will is free; that what is good for me is also good in itself. (1882, 110)

In fact, Nietzsche was pitching it here a bit stronger than he really meant. It was not so much that everything is false, but that everything could be false. Because the epistemic principles work and because we believe in them – we would, wouldn't we! – it does not follow that they are true. We have cut ourselves off from an external touchstone. This worry is shared by the contemporary Christian philosopher Alvin Plantinga, although he does not mention Nietzsche as a predecessor, whether from ignorance of the history of philosophy or because he prefers not to acknowledge the priority of one so hostile to religion as Friedrich Nietzsche, a man who declared and rejoiced in the "Death of God." Like Nietzsche, Plantinga argues that Darwinian evolution cares nothing for truth, only for survival and reproductive success. To use a memorable phrase of the philosopher Pat Churchland, Darwinism is the science of the "four Fs": fighting, fleeing, feeding, and reproduction. There is nothing here about knowledge and truth and objectivity. All that our biology need do is tell us what we need to believe to survive and reproduce – and this information, so long as it is effective, could as easily be quite false. Plantinga somewhat cutely labels this "Darwin's Doubt," because it was even expressed by a worried Darwin himself. "With me the horrid doubt always arises whether the convictions of man's mind, which have been developed from the mind of the lower animals, are of any value or are at all trustworthy. Would anyone trust in the convictions of a monkey's mind, if there are any convictions in such a mind?" (Plantinga 1993, 219, quoting F. Darwin 1887, I, 315–16). As a matter of fact, Darwin immediately excused himself as a reliable authority on such philosophical questions, but this somewhat awkward point goes unmentioned. As does the fact that Darwin was explaining why he saw purpose in life, something one might have thought Plantinga would have welcomed.

Making the case in his own right, Plantinga tells the story of a posh dinner in an Oxford college. Apparently, Richard Dawkins spoke up for atheism before the philosopher A. J. Ayer. (A classic case of coals to Newcastle if ever there was one.) On the basis of this, Plantinga draws the Nietzschean conclusion: perhaps we are living in a dream world, and none of our thoughts tells us a thing about the real world. The beliefs of such beings as this

> might be like a sort of decoration that isn't involved in the causal chain leading to action. Their waking beliefs might be no more causally efficacious, with respect to their behaviour, than our dream beliefs are with respect to ours. This could go by way of pleiotropy: genes that code for traits important to survival also code for consciousness and belief; but the latter don't figure into the etiology of action. It *could* be that one of

these creatures believes that he is at that elegant, bibulous Oxford dinner, when in fact he is slogging his way through some primeval swamp, desperately fighting off hungry crocodiles. (Plantinga 1993, 223–24)

Obviously, if everything that we believe is false, then clearly everything we believe about evolution could be false – natural selection, tree of life, and so forth. And so everything collapses into a contradictory mess, the whole of Darwinian epistemology ends in a *reductio ad absurdum*. Since our theory of knowledge embraces indifferently the true and the false, so long as it is expedient, we are in deep trouble.

How seriously should we take this line of argumentation? How devastating is it to this kind of evolutionary epistemology? Grant what we have already accepted, namely that there are times when organisms and their characteristics will be out of adaptive focus. Genetic drift is a case in point, as is that phenomenon mentioned by Plantinga, "pleiotropy," where a single gene controls two different characteristics and a non-adaptive feature piggybacks on an adaptive feature. But because something logically could happen, like the crocodile tale, it does not follow that it has the slightest basis in contemporary biological actuality as dictated by Darwinian theory. What Plantinga posits is simply absurd. Almost by definition, drift has minor effects – effects so minor that they can slip under selection. Thinking that you are matching eminent philosophers drink for drink may be the way to conjure up pink rats, but it is not the way to fight off crocodiles. At the very least, if you are going to mimic Crocodile Dundee, you need strength and cunning and lightning-fast reactions to threat. If drift or other non-adaptive-feature-producing mechanisms are getting in the way of selection for the needed attributes, you are indeed in deep trouble. For this reason, nothing in Darwinian theory suggests that adaptive focus is a luxury, something that might or might not obtain. If it is needed, either you have it or you are wiped out.

Take the discussion to a more serious level. Nietzsche and Plantinga do have a point. There is nothing in Darwinian theory suggesting that it is never the case that (as it were) selection deceives us for our own good. Perhaps we do work better when we believe things that are not true. A Humean readily concedes this. Take causal connection. A child puts its hand in the fire. It burns, the child feels pain and screams, and pulls its hand out quickly before any great damage is done. From now on, the child avoids the fire with great care. Why? Because the fire has the "power" to cause burning. But as Hume pointed out, there is no power there, just constant conjunction which sets up in us an erroneous belief that there is such a power. And there is good reason for this. If our psychology, something we are now interpreting in Darwinian terms, did not set up such a belief, we would be in big trouble. To our very great detriment, we would never learn to fear and thus avoid the fire. Selection is deceiving us because we are better off that way. The truth may set you free. It could also kill you.

But this is only half of the story. Systematic misconceptions don't occur randomly. They occur when we need them, and the big point – the point without which we could not have this discussion – is that we can work out why such misconceptions occur and why it is that they come from selection. Think of the other side of the coin. There are lots of cases where no misconceptions do occur or can occur – at least, none where it would be in selection's interests that we be properly informed. Falling trees hurt; drinking arsenic kills; beautiful naked people (with provisos about sex and orientation) are a sexual turn-on; roses really are red. These are not and cannot be cases of misconception. Selection is not about to kid us on such a topic as the hurtfulness of falling trees. The world is not crazy. That is why Plantinga's example of fighting crocodiles while you think you are in Oxford uttering wise thoughts is so unconvincing. From a Darwinian perspective it is truly crazy. The deceptions of selection make sense, just as do the non-deceptions. We cannot be deceived all of the time. And in fact, the Darwinian hand is even stronger than this. Generally, evolution leads to true conclusions. So we can and do use these to ferret out the cases where we are deceived, those instances where selection does deceive us. Hume did not have to leave the human race to do his philosophy.

Like René Descartes in his *Meditations*, let us push our doubts to the extreme. Have we ruled out the possibility of some kind of overall systematic deception? Is it not possible that we are like the prisoners in Plato's cave in the *Republic*, thinking that shadows on the wall are the real thing? Suppose, hypothesizes Plantinga, we work in a factory where everything seems red. The non-factory worker, the person outside, knows that it is all a question of filters. The workers within are ignorant and deceived. Unbeknownst to them, they are wearing spectacles which color their perceptions. In truth nothing is red. But it is only on leaving that you can discover this. Within the factory, the worker has no genuine touchstone by which to make a judgment. It is not a matter of taking a bit more care. Within the factory, all is illusion. Could we humans not be all in the same situation? Is it not possible, to push Nietzsche's point, that everything in the world/factory is false/red-seeming-but-not-really? (Not that everything is, but that everything could be.)

I think at this point the Darwinian has to cry *cave*! If you insist, then probably you can never escape from the ultimate doubt that it is all a cosmic joke and that logically everything could be a fraud. In a way, you have a situation parallel to the notorious brain-in-a-vat dilemma. How can you be sure that you really are up and walking about and not a brain in a vat of chemicals, all wired up to have the right sensations and illusions of reality? Some philosophers, notably Hilary Putnam, think the brain-in-a-vat case collapses into absurdity. I am not sure. What I am sure of is that whether or not it is absurd, it doesn't really matter. Life goes on, whether for real or not. And alternatives are not much more convincing. Descartes thought he could escape from his systematic doubt first by recognizing that one claim – *cogito ergo sum* – could not be doubted. Or rather, to doubt it is to reaffirm it. From there he wanted to get to God, who would guarantee the truth of everything, or at least of our clear and distinct ideas. But as virtually everyone has pointed out, wanting to get to God is not the same as getting to God, and most think that having dug himself into the hole of doubt, he never gets out. Plantinga thinks that our knowledge of God and his truth-affirming nature is basic, and that it is immune from outside doubt or skepticism. To which one can only reply that that may be his opinion but that the rest of us smell self-deception. What kind of truth is it that a supposedly good God hides from honest doubters?

Theories of knowledge

Coming to the end of the discussion, however, let us concede or rather acknowledge a pair of related very important points. There is this much truth in the critics' arguments that the supporter of this kind of Darwinian epistemology is being pushed to a coherence theory of truth, where what counts is that everything hang together, rather than a correspondence theory of truth, where what counts is that one's ideas correspond to objective reality. Lorenz and Popper and others are wrong to think that Darwinism solves the Kantian dilemma of the unknowable thing-in-itself. At some level, as Thomas Kuhn, not to mention the pragmatist Richard Rorty (1979), argued, you are trapped in your own thoughts and can never truly get out. This is not necessarily worrisome and does not mean that within the system you cannot make judgments about what is real and what unreal, what true and what false. But it is always within the system.

Hilary Putnam refers to this kind of thinking as "internal realism":

> One of these perspectives [on realism] is the perspective of metaphysical realism. On this perspective, the world consists of some fixed totality of mind-independent objects. There is exactly one true and complete description of "the way the world is". Truth involves some sort of correspondence relation between worlds or thought-signs and external things and sets of things. I shall call this perspective the *externalist* perspective, because its favorite point of view is the God's Eye point of view.
>
> The perspective I shall defend has no unambiguous name. It is a late arrival in the history of philosophy, and even today it keeps being confused with other points of view of a quite different sort. I shall refer to it as the *internalist* perspective, because it is characteristic of this view to hold that *what objects does the world consist of?* is a question that it only makes sense to ask *within* a theory or description. Many "internalist" philosophers, though not all, hold further that there is more than one "true" theory or description of the world. "Truth," in an internalist view is some sort of (idealized) rational acceptability – some sort of ideal coherence of our beliefs with each other and with our experiences *as those experiences are themselves represented in our belief system* – and not correspondence with mind-independent "states of affairs." There is no God's Eye point of view that we can know or usefully imagine; there are only various points of view of actual persons reflecting various interests and purposes that their descriptions and theories subserve. (Putnam 1981, 49–50)

As noted, the theory of truth embraced here is the coherence theory – the aim is to get things to hang together – although this does not preclude correspondence talk within the system.

I should add that, going on to the second point, and pushing the argument further, if one accepts some of the claims made earlier about metaphor and the nature of science, one may well find oneself with those internalist philosophers who do accept that there can be more than one true theory or description of the world (Ruse 1999). Assume for the sake of argument that in this world or in any world, a kind of Kantian assumption about rationality holds – that the principles of logic and mathematics and of good science will be the same. In other words, that evolution will throw up basically the same innate dispositions. I am not sure that this is true, not because I doubt the ubiquity of Darwinian selection – I very much expect that this does hold generally – but because of such things as non-Euclidean geometries and the like. But making the concession does not affect the main argument. You have still got the metaphors to reckon with, and if anything is deeply cultural it is the metaphors of science. Division of labor is, literally, right out of Adam Smith in the eighteenth century and the first flowering of the British Industrial Revolution. An arms race is one of the defining metaphors of the twentieth century. And the tree of life is about as deeply rooted (!) in the Judeo-Christian West as it is possible for anything to be. Perhaps some of the metaphors of Darwinian biology rise above the immediately cultural. The notion of design perhaps comes with the possibility of rationality. It is hard to imagine a rational being without any conception of artifacts. But after that it is specific culture all of the way. Suppose, however, one had a society where there was no Industrial Revolution and no competition and no building up of weapons systems and so forth. It is sometimes said that Japan fits this non-Western mould, although I am not really sure. Certainly it has been in the Western orbit in the last century or so. But suppose one did have such a society. Would it be possible for it to have a scientific theory of origins and if so what form would it take? One very much doubts that natural selection would play a big role in such a theory, especially if the society was not agricultural but hunter-gatherer in some sense – say getting food from the sea or from a widely growing and available herb. Indeed it is doubtful, if one did not already have a religious or similar story of the past, whether one would have a theory of origins as such. A major reason why Plato and Aristotle were not into evolution was simply because they did not see the need for it. We have the Jewish sacred writings to thank or blame for that obsession.

Now note what is being said here, or rather what is not being said here. No one is saying that anything goes or that any metaphor is as good as the next one. That is simply not true. We are confining our discussion to science – we know already that different societies have different (incompatible) creation stories. The point about science is that it does have standards, epistemic-value principles, and any science will be judged by them. Darwin's theory works on its merits, not because we all accept the joys and benefits of industrialism. I can quite well imagine that at a personal level Wilson has grave doubts about the divisions of labor – Japanese car manufacturers that break down the barriers and give everyone shared work seem to do better than American car manufacturers working on a strict division pattern – but he still finds the metaphor very useful when he applies it to ants. Unlike humans, ants do not get bored doing the same thing, day in and day out. Nor is anyone saying that a rival theory to (say) Darwinism is going to contradict Darwinism. No one is saying that in another place or at another time, Creation Science or Intelligent Design Theory might be true. No one is saying that the Mormons could have it right and the native people of North America might be the lost tribes of Israel rather than nomads who crossed the Bering Strait. The point is that any rival theory is going to look at things in a different way, perhaps not even in a way that we would recognize as evolutionary. If I could tell you in what way, I would, but because I am part of our society I cannot.

Perhaps I am out on a limb and sawing it off. It would not be the first time! Let me simply say, and this is a good point on which to end the chapter, that if you take seriously the idea that science is a human construction – and that humans are biological beings immersed in culture – you should be very careful about any theory of science that does not take seriously human nature, its powers, and its limits.

6   Morality

"What should I do?" Let us turn now to the second of the great questions of philosophy. Does evolutionary theory, does Darwinian evolutionary theory, throw any light on this topic? Many people today think that it does, but how far the light actually penetrates is still a matter of great controversy. As with the problem of knowledge, it makes most sense to go back and start with Charles Darwin himself.

Darwin on morality

Thanks to an extended discussion in the *Descent of Man*, Darwin had far more to say on the topic of morality and behavior than he had had to say on knowledge and its foundations. Moreover, his thinking on the issues takes us right to the heart of one of the most contested issues in contemporary evolutionary theory. But start as before with the fact that, although knowledgeable, Darwin was not a philosopher with a philosopher's questions. He was a scientist with a scientist's questions, and philosophy would be tackled (if at all) only tangentially. We see this at once in Darwin's treatment of morality. Philosophers distinguish between two major issues that must be addressed when dealing with moral thought and behavior. What should I do? Why should I do what I should do? These two branches of the subject are usually referred to as "normative (or substantive) ethics" and "metaethics," the first to do with directions and the second with foundations. By way of example in Christian ethics one finds normative questions along the lines of "Love your neighbor as yourself," and generally a metaethical answer in terms of God's will, "You should do that which God wants." Of course giving answers is only the beginning of inquiry. Who is one's neighbor? Why should one obey God? But this is as it may be. I am using Christianity as an example of how it all works, not as solution which should (or should not) be accepted.

The important point for us here is that Darwin certainly was not going to present his discussion in terms of this dichotomy. We may think that Darwin has things to say about both levels and we may think what he has to say has relevance today, but we are going to be on our own in finding these out. In fact, at the normative level, Darwin was reasonably informative, though we need to take note that his discussion does not focus on the offering prescriptions for good behavior so much as assume these, trying to explain them in the light of evolutionary biology. The popular mid-Victorian ethical theory was utilitarianism in some form or another. The Greatest Happiness Principle: "You ought to promote the greatest happiness, and the least unhappiness, for the greatest number of people." John Stuart Mill's famous essay on the topic appeared in 1863, although historically the theory dates at least to the eighteenth century and Darwin's grandfather Erasmus endorsed a version. The progress of evolution "is analogous to the improving excellence observable in every part of the creation ... such as the progressive increase of the wisdom and happiness of its inhabitants" (E. Darwin 1794–96, 509). Charles Darwin accepted this philosophy, why would he not, but gave it a bit of a biological twist. "The term, general good, may be defined as the term by which the greatest possible number of individuals can be reared in full vigor and health, with all their faculties perfect, under the conditions to which they are exposed" (C. Darwin 1871, I, 98).

To be honest, however, Darwin was not really tremendously reflective on what all of this might mean in practical terms, although it is fairly easy to infer that his values were those of an upper-middle-class Englishman, of a liberal persuasion. He was violently against slavery, and unlike many of his countrymen supported the North in the American Civil War. (A lot of the British support for the South came from the close connection between the cotton fields of the breakaway states and the cotton mills of industrial Lancashire.) However, equally, he was in favor of capitalism and we know that he regarded with horror proposals that working men should be allowed to form unions. This violates sound economics as well as intrudes on owners' liberties. Although by 1871 (the year of the *Descent*) Darwin's religious beliefs had moved over to agnosticism, like others (including the archetypal critic of Christianity, Thomas Henry Huxley) he endorsed Bible-based religious instruction on the grounds that this was an essential part of moral training.

What Darwin did see was that, at the biological level, morality demands an element of sociability. We have got to have a feeling that we can and want to get on with our fellows.

> It has often been assumed that animals were in the first place rendered social, and that they feel as a consequence uncomfortable when separated from each other, and comfortable whilst together; but it is a more probable view that these sensations were first developed, in order that those animals which would profit by living in society, should be induced to live together, in the same manner as the sense of hunger and the pleasure of eating were, no doubt, first acquired in order to induce animals to eat. The feeling of pleasure from society is probably an extension of the parental or filial affections, since the social instinct seems to be developed by the young remaining for a long time with their parents; and this extension may be attributed in part to habit, but chiefly to natural selection. With those animals which were benefited by living in close association, the individuals which took the greatest pleasure in society would best escape various dangers, whilst those that cared least for their comrades, and lived solitary, would perish in greater numbers. With respect to the origin of the parental and filial affections, which apparently lie at the base of the social instincts, we know not the steps by which they have been gained; but we may infer that it has been to a large extent through natural selection. (C. Darwin 1871, I, 80)

Note, an important point to be picked up in a moment, that not only is it natural selection that causes all of this, but that the key lies in "parental and filial affections."

We next get a move that shows that, although he was no professional philosopher, we should not underestimate Charles Darwin. Morality, he sees, is not just a matter of unreflective action, but of conscious deliberation leading to action. We have first-order desires, and then second-order reflections on the desires and the actions they promote. This is or leads to a conscience. When he was developing his theory in the late 1830s, Darwin had read Hume on animal reasoning and always saw a continuity between the ape and the human. But he did want to emphasize that it is this ability to second-order reflect that distinguishes a moral animal like a human from a mere brute. "The following proposition seems to me in a high degree probable – namely, that any animal whatever, endowed with well-marked social instincts, would inevitably acquire a moral sense or conscience, as soon as its intellectual powers had become as well, or nearly as well developed, as in man" (C. Darwin 1871, I, 71–72). All of this worrying about what you should do rather than just getting on with it sounds more Germanic than British. After all, Hume had calmly decreed that "reason is a slave of the passions"! Kant on the other hand put a good will above any ends, good, bad, or indifferent. Confirming our suspicions, we find that before writing the *Descent*, Darwin did read Kant's *Metaphysics of Morals* and in *Descent* actually quotes one of the more purple passages in that work. "Duty! Wondrous thought, that workest neither by fond insinuation, flattery, nor by any threat, but merely by holding up thy naked law in the soul, and so extorting for thyself always reverence, if not always obedience; before whom all appetites are dumb, however secretly they rebel; whence thy original?" (I, 70). In truth, however, Darwin was never really on that track. For Kant, morality had a kind of necessity, stemming from the conditions required for rational beings living together. If there are rational beings on Andromeda, they will think and behave like late eighteenth-century Germans living on the far reaches of the Baltic, revealing their Pietist childhoods. Darwin to the contrary bluntly asserted that if things had gone otherwise, we might think that killing each other is the highest moral duty.

> It may be well first to premise that I do not wish to maintain that any strictly social animal, if its intellectual faculties were to become as active and as highly developed as in man, would acquire exactly the same moral sense as ours. In the same manner as various animals have some sense of beauty, though they admire widely different objects, so they might have a sense of right and wrong, though led by it to follow widely different lines of conduct. If, for instance, to take an extreme case, men were reared under precisely the same conditions as hive-bees, there can hardly be a doubt that our unmarried females would, like the worker-bees, think it a sacred duty to kill their brothers, and mothers would strive to kill their fertile daughters; and no one would think of interfering ... The one course ought to have been followed, and the other ought not; the one would have been right and the other wrong. (C. Darwin 1871, I, 73–74)

Although I am not sure of the extent to which there was a direct input from Hume to Darwin on the topic of morality – I am much more sure about the origins of religion, where there is convincing documentary evidence – we can certainly say that Darwin stood firmly in the British empiricist tradition when it came to morality. The key notion (used by Adam Smith and Edmund Burke as well as Hume) was always that of "sympathy" – a kind of moral feeling that one has for others – and this was central to Darwin's thinking also. "The aid which we feel impelled to give to the helpless is mainly an incidental result of the instinct of sympathy, which was originally acquired as part of the social instincts, but subsequently rendered … more tender and more widely diffused" (I, 86). And: "Nor could we check our sympathy, even at the urging of hard reason, without deterioration in the noblest part of our nature. The surgeon may harden himself whilst performing an operation, for he knows that he is acting for the good of his patient" (I, 169). It hardly takes a daring inference to conclude that Darwin's metaethics, his justification for normative prescriptions, was thoroughly naturalistic. Any foundations have to be in terms of human nature and this is our biological human nature. What these foundations might be, however, or if indeed there are any foundations at all, is not a question raised by Darwin.

Levels of selection

As intimated above, before moving on to the thinking of others, we must first pause and turn more directly to the workings of Darwinian evolutionary biology. If you think about it, on the face of things, morality is profoundly non-Darwinian and should never have appeared. To hell with being nice to others! There is a struggle for existence between organisms and victory is all important. Win or die! This was certainly the conclusion of others, not the least of whom (in historical importance) was Adolf Hitler.

> All great cultures of the past perished only because the originally creative race died out from blood poisoning.
>
> The ultimate cause of such a decline was their forgetting that all culture depends on men and not conversely; hence that to preserve a certain culture the man who creates it must be preserved. This preservation is bound up with the rigid law of necessity and the right to victory of the best and stronger in this world.
>
> Those who want to live, let them fight, and those who do not want to fight in this world of eternal struggle do not deserve to live. (Hitler 1925, ch. 11)

We have already seen the pieces to Darwin's answer to all of this. Being nice to others pays dividends to oneself. The social animal gives out care and sympathy, but can expect it back in return. There is a struggle for existence, but success in the struggle can be more subtle than simply beating the other into a bloody pulp. It can involve cooperation, because half a cake is a lot better than none at all. But this now raises the question of how exactly natural selection brings all of this about, and here we do start to dig right down into some of the most basic and important questions in the whole of evolutionary theory. Introducing the term "altruism" in the sense used by modern biologists, meaning help given to others at your own immediate (reproductive) cost, and recognizing that this is a metaphor and not necessarily "altruism" in the literal sense of helping others because you know that you should, there are two ways in which selection might promote altruism. In the first case, it might come about because the "unit of selection" is the group to which the altruist belongs. In other words, although the altruist pays, the group gains because of the altruist and selection favors this group over others. In this case, where we have "group selection" at work, the group is rather like the organic body and the individual is like a part (a heart or a lung) where what the part does is lost within the functioning of the whole. In the second case, you insist that the "unit of selection" is the individual organism. In this case, although the altruist puts out, it must be with the expectation that there will be a payoff later. Overall, the gain is to the individual altruist. In other words, altruism is enlightened self-interest.

Darwin was fascinated by this issue, virtually from the moment that he discovered natural selection (Ruse 1980). Most particularly, he was fascinated by the challenge thrown up by the social insects. Everybody knew – certainly every schoolboy who had had to translate Virgil's *Georgics* knew – about the honeybee and about the sacrifice that the workers make for the good of the hive. Many wrote on and around these topics in the eighteenth and early nineteenth centuries and, of course, Darwin came from rural England, where bee-keeping, a hive or two at the bottom of the garden, was commonplace. The social insects put the problem of altruism up front because the workers, members of the group devoting all of their energies to others, are sterile. It is the queens and the males who do all of the reproducing. How can this be? Really there are two questions here (Dixon 2008). First, how can sterility get passed on? How can you inherit it? Second, what adaptive advantage can sterility have? Surely being a reproducer is far better? A Lamarckian has trouble with the first question, because sterility is hardly an acquired characteristic that is transmitted. It is one thing to study the Jews and see if eventually penises are produced naturally without foreskins. It is another thing to study eunuchs and to see if eventually they are produced naturally. However, Darwin (1859) saw how it could happen so long as one thinks in terms of families. Sterility in the social insects is not something they are born with, but rather something acquired during growth. In other words, in our terms there seems to be a genetic predisposition to sterility given the right circumstances. How could this be inherited? Simply by being passed on by those family members who do reproduce. They have the predisposition and they are selected, and they are not sterile. As Darwin pointed out, the situation is exactly what you find in animal breeding. You have a fine carcass from an ox and so you go back to the parent stock (or any close relative stock) and breed again. The relatives pass on the features, and if indeed instead of well-marbled meat it is a propensity to work hard and not to reproduce, then so be it. Nature is satisfied.

But how does selection mimic the butcher? How does selection actually favor the sterile worker? The obvious answer is group selection, but Darwin was always somewhere between very uncomfortable and outright opposed to the idea. (He did not have the term "group selection," but he knew very well what it was all about.) Right from the beginning he saw the struggle as being an individual matter. "Hence, as more individuals are produced than can possibly survive, there must in every case be a struggle for existence, either one individual with another of the same species, or with the individuals of distinct species, or with the physical conditions of life" (C. Darwin 1859, 63). Why did he feel so strongly on this matter? In part, I strongly suspect, it was a consequence of his overall politico-economic philosophy. He was a Malthusian, and right after the passage just quoted, went on to say that he was endorsing the philosophy of that particular thinker. Darwin really did think of life as being a battle between individuals. This contrasts nicely with the co-discoverer of natural selection, Alfred Russel Wallace, who was a lifelong socialist and who happily embraced group selection. In part, perhaps, Darwin thought as he did because he wanted to be consistent. After all, sexual selection is virtually by definition an individual-selection mechanism – a battle between organisms in the group – and so it is natural to think of natural selection the same way. As we shall learn, Wallace was a lot less keen on sexual selection. In part, although I think it was only in the 1860s when he had the matter out with Wallace and became fully aware of the underlying issues that he would have been able to articulate this objection, Darwin was wary of group selection because of the matter that sticks in the craw of evolutionists today. Group selection is too open to cheating. If you have a collection of group altruists, then a variant cheater who does nothing for others is going to be at a selective advantage, and its offspring will spread at the expense of the altruists.

How then can individual selection promote altruism? Today's evolutionists have proposed two main mechanisms. "Reciprocal altruism," where help is given in anticipation of help received (Trivers 1971). "You scratch my back and I'll scratch yours." And something called "kin selection," where altruists may not reproduce but by helping close relatives to reproduce they thereby get to pass on more copies of their own units of heredity, their own genes (Hamilton 1964). The latter is particularly important when it comes to explaining why, in the social insects, workers devote all of their efforts to raising the fertile offspring of their mother. It is a question of benefits outweighing costs. If one is in a place well defended and with food supplies, an individual's best reproductive strategy ("fortress defense") might be to raise related organisms (albeit not as related as one's own offspring) rather than venture afield with the risk of injury and death (and hence no offspring). In the hymenoptera, another strategy ("life insurance") seems important. Here it is in an individual's reproductive interests to stay with the nest because the coverage of the young (even if not yours) is better (thanks to the nest mates) than if you go it alone (and perhaps die before raising your young).

It must be stressed that kin selection in particular has led to some stunning, confirmed predictions. In the hymenoptera, females are born of fertilized eggs whereas males from unfertilized eggs. The result is that full sisters are three-quarters related and sisters to brothers a quarter-related (unlike humans, say, which are half-related for both sisters and brothers). However, if the mother (the queen) had multiple fertilizations, then the relationship between females starts to drop (because they have different fathers). Hence, in single-mated nests one expects the workers to skew sex ratios toward female siblings (because you are raising more of your own genes that way) and in multiple-mated nests one does not expect this skew. The prediction holds repeatedly! Darwin did not have Mendelian genetics, so he could not devise explanations like these. But he did see – remember how I highlighted the point about sociality starting in families – that, even if sterile, close relationships would enable you to pass on your own heredity. To be honest, Darwin varied between seeing each individual as a unit and the hive or the nest as a unit, but even in the latter case, the important point is that the hive or nest members are close relatives. You cannot have a selection unit of unrelated individuals.

This is all basically in the *Origin*. As noted, Darwin and Wallace then went on to discuss these issues in the 1860s. Hence, by the time Darwin came to write the *Descent*, he was fully sensitized to the topic, as of course he needed to be because moral behavior does involve altruism in an extreme form. (Note that the point of Darwin's discussion was to show that biologically we are altruists, but that the way in which we humans achieve our biological ends is through literal, ethical altruism. Doing good serves our reproductive goals. Not always, of course, but on average.) Does this mean that Darwin stuck to an individual-selection stance throughout, or did he weaken (say, rather, modify) and allow that, at the human level, group selection might take over? He could certainly have used the excuse that humans with their powers of language and of thought so change the formula that all previous bets are off. (This is the position of Richerson and Boyd 2005.) And he recognized that there are some very knotty issues that do seem to call for group selection. "It is extremely doubtful whether the offspring of the more sympathetic and benevolent parents, or of those who were the most faithful to their comrades, would be reared in greater numbers than the children of selfish and treacherous parents of the same tribe" (C. Darwin 1871, I, 163). In part, he got around the problem by appealing to (individual-selection promoting) reciprocal altruism. "In the first place, as the reasoning powers and foresight of the members became improved, each man would soon learn that if he aided his fellow-men, he would commonly receive aid in return. From this low motive he might acquire the habit of aiding his fellows; and the habit of performing benevolent actions certainly strengthens the feeling of sympathy which gives the first impulse to benevolent actions. Habits, moreover, followed during many generations probably tend to be inherited"

(163–64). (Note the Lamarckism, something that Darwin always accepted along with selection.) But then he did seem to bite the bullet and go in a group selection way.

> It must not be forgotten that although a high standard of morality gives but a slight or no advantage to each individual man and his children over the other men of the same tribe, yet that an advancement in the standard of morality and an increase in the number of well-endowed men will certainly give an immense advantage to one tribe over another. There can be no doubt that a tribe including many members who, from possessing in a high degree the spirit of patriotism, fidelity, obedience, courage, and sympathy, were always ready to give aid to each other and to sacrifice themselves for the common good, would be victorious over most other tribes; and this would be natural selection. At all times throughout the world tribes have supplanted other tribes; and as morality is one element in their success, the standard of morality and the number of well-endowed men will thus everywhere tend to rise and increase. (166)

I am not, however, convinced that this is quite as definitive as it seems. Darwin is talking about the tribe, and he makes it clear that there is going to be a lot of intermarriage in such a group. In other words, there will be a lot of blood relationships. And in a letter to one of his sons (George, a physicist) later in the decade, he affirms what we might expect, namely that he is thinking of the tribe as akin to a hymenopteran nest, where we do get such close relationships.

> To G. H. Darwin 27 April [1876]
> Down Beckenham Kent
> Ap. 27 th
> My dear George
> I send "Mind" – it seems an excellent periodical – Sidgwicks article has interested me much. – It is wonderfully clear & makes me feel what a muddle-headed man I am. – I do not agree on one point, however, with him. He speaks of moral men arising in a tribe, accidentally, i.e. by so-called spontaneous variation; but I have endeavoured to show that such men are created by love of glory, approbation &c &c. – However they appear the tribe as a tribe will be successful in the battle of life, like a hive of bees or nest of ants.
> We are off to London directly, but I am rather bad.

> Leonard comes home on May 10 th !! Plans changed. (Darwin Correspondence Project Database, www.darwinproject.ac.uk/entry-10478 [letter no. 10478, unpublished])

In response to a young supporter, George John Romanes, who wrote pressing on him a group-selection perspective, Darwin also demurred (Romanes 1895, 173).

Social Darwinism

Let us move on now from Darwin. We saw in the discussion of epistemology that there were two main approaches. One worked more at the cultural level and the other was the more literal approach, working from the supposed nature of the human mind. We find something very similar in the case of ethics. The first, more traditionalist approach is often today known as "Social Darwinism" – a term incidentally invented in the 1940s (in other words unknown to the people whom it purports to represent) and with a somewhat unfortunate reputation as referring to a system or approach that is (from a normative perspective) pretty vile and, one can say with some relief, no longer fashionable. As is so often the case, history is a little more complicated. Herbert Spencer is the first and still the main representative of this approach. Essentially (so the story goes) he saw human societies in the way he saw animal and plant societies, namely as engaged in a bloody struggle for existence. In the human realm, this translated out into a form of *laissez-faire* economics, where the state stands aside and naked competition determines the winners and losers. Drawing on this, Spencer then argued that it is all a very good thing and what one ought to do. In other words, his normative ethics came straight out of the Darwinian selection processes. That there is some truth in all of this cannot be denied.

> We must call those spurious philanthropists, who, to prevent present misery, would entail greater misery upon future generations. All defenders of a Poor Law must, however, be classed among such. That rigorous necessity which, when allowed to act on them, becomes so sharp a spur to the lazy and so strong a bridle to the random, these pauper's friends would repeal, because of the wailing it here and there produces. Blind to the fact that under the natural order of things, society is constantly excreting its unhealthy, imbecile, slow, vacillating, faithless

> members, these unthinking, though well-meaning, men advocate an
> interference which not only stops the purifying process but even increases
> the vitiation – absolutely encourages the multiplication of the reckless
> and incompetent by offering them an unfailing provision, and *discourages*
> the multiplication of the competent and provident by heightening the
> prospective difficulty of maintaining a family. (H. Spencer 1851, 323–24)

However, before we rush to judgment, we should note that sentiments like
these were held before Spencer became an evolutionist; in any case, as we
know already, although Spencer was an independent discoverer of natural
selection, for him the main mechanism of change was always Lamarckism.
The connections between this kind of thinking and Darwinian evolutionary biology are loose, to say the least. Moreover we need to put this kind of
talk in context. They are the writings of one who stood outside the British
establishment, which passed restrictive laws (like the Corn Laws) for its
own good, and which took little heed of the needs and achievements of
the creative and hardworking, especially those whose thought, daring,
and labor had brought about the Industrial Revolution. The language and
sentiments are harsh, but they expressed views held by many. In any case,
Spencer was more eager to argue against the value of militancy and the
building up of weapon stocks, for he saw this as inimical to free trade. He
balanced *laissez-faire* sentiments with a remarkably pacifistic foreign policy (Dixon 2008; Richards 1987).

This is not to deny that, after he became an evolutionist, Spencer was
not slow to link his social thought with his biological thought, and that
others were ready to follow in his path. A good example, one that shows
that there is a lot more to traditional evolutionary ethics than what is usually thought to fall under the heading of Social Darwinism, is the Russian
anarchist exile, Prince Petr Kropotkin. An unabashed group selectionist,
he saw all organisms including humans as having developed a propensity
to what he called "mutual aid" and held that this is and should be the basis
of morality.

> Two aspects of animal life impressed me most during the journeys which
> I made in my youth in Eastern Siberia and Northern Manchuria. One of
> them was the extreme severity of the struggle for existence which most
> species of animals have to carry on against an inclement Nature; the
> enormous destruction of life which periodically results from natural

agencies; and the consequent paucity of life over the vast territory which fell under my observation. And the other was, that even in those few spots where animal life teemed in abundance, I failed to find – although I was eagerly looking for it – that bitter struggle for the means of existence, among animals belonging to the same species, which was considered by most Darwinists (though not always by Darwin himself) as the dominant characteristic of struggle for life, and the main factor of evolution. (Kropotkin 1902, vi–viii)

What Kropotkin did see was one animal helping another. On the one hand, there was the appallingly harsh environment.

On the other hand, wherever I saw animal life in abundance, as, for instance, on the lakes where scores of species and millions of individuals came together to rear their progeny; in the colonies of rodents; in the migrations of birds which took place at that time on a truly American scale along the Usuri; and especially in a migration of fallow-deer which I witnessed on the Amur, and during which scores of thousands of these intelligent animals came together from an immense territory, flying before the coming deep snow, in order to cross the Amur where it is narrowest – in all these scenes of animal life which passed before my eyes, I saw Mutual Aid and Mutual Support carried on to an extent which made me suspect in it a feature of the greatest importance for the maintenance of life, the preservation of each species, and its further evolution. (viii)

Clearly there was more to Social Darwinism than simple competition. Or if you prefer to keep the term "Social Darwinism" for competitive philoso-phies, then we can say that there was more to traditional ways of using biology as a guide to moral behavior than simple exhortations to compe-tition. Not, of course, that this is to deny that there were those who did embrace a crude vision of life's struggles, seeing these both as the norm and as in some sense to be admired if not cherished. In his story of dogs battling for supremacy in the Far North, *The Call of the Wild*, the American novelist Jack London pitched things strongly.

There was no hope for him. Buck was inexorable. Mercy was a thing reserved for gentler climes. He manoeuvred for the final rush ... Then Buck sprang in and out; but while he was in, shoulder had at last squarely met shoulder. The dark circle became a dot on the moon-flooded snow as Spitz disappeared from view. Buck stood and looked on, the successful champion, the dominant primordial beast who had made his kill and found it good. (London 1903, ch. 3)

And so on and so forth. Somewhat more ominously and dangerously, escaping the bounds of adolescent fiction, these sorts of sentiments found their way into military manuals, including the highly influential *Germany and the Next War* by General Friedrich von Bernhardi, sometime member of the German High Command. It is morally good and acceptable that Darwinism endorses war. "Struggle is ... a universal law of Nature, and the instinct of self-preservation which leads to struggle is acknowledged to be a natural condition of existence. 'Man is a fighter'" (von Bernhardi 1912, 13). And: "might gives the right to occupy or to conquer. Might is at once the supreme right, and the dispute as to what is right is decided by the arbitration of war. War gives a biologically just decision, since its decisions rest on the very nature of things" (15). Hence: "It may be that a growing people cannot win colonies from uncivilized races, and yet the State wishes to retain the surplus population which the mother-country can no longer feed. Then the only course left is to acquire the necessary territory by war" (15). It is little surprise that during and after the First World War, many people thought that any application of Darwin's thinking to humankind was necessarily dangerous, and worse. There is good reason to think that the chief anti-evolution motivation of the Great Commoner, three-times presidential candidate William Jennings Bryan, the man who led the prosecution in the Scopes Monkey Trial in Tennessee in 1925, when a young school teacher was put on trial for teaching evolution, stemmed from precisely these factors. He hated the militarism that led to the war and he fingered Darwinism as a major culprit (Larson 1997).

What about Hitler? In certain circles today – notably those of the anti-evolution biblical literalists, the Creationists, and their successors the Intelligent Design Theorists – there is an awful lot of enthusiasm (if that is the right word) for the Darwin-led-to-Hitler-led-to-the-Holocaust thesis. (For example, Weikart 2004.) And with their case bolstered by sentiments like some of the quotations given earlier, it does seem as though the critics have a point. However, as always, things are a bit more complex than that. There was no one thing that led to Hitler and the Nazis (Kershaw 1999). Putting it all down to Darwin is as historically crude as (another popular thesis) putting it all down to Christianity, meaning the anti-Semitism of Martin Luther. Hitler, to put it mildly, was not a well-educated man, and he had picked up half-digested ideas from all over, including in particular huge amounts of dangerous nonsense about the Germans as a *Volk* and the ideology he saw manifested in Wagner's operas. (How anyone can see the *Ring* as a triumph of the German spirit is a topic ripe for a thesis in itself.) Moreover, if you look at the supposedly Darwinian passages in context, you see that Hitler's real obsession is with racial purity, and this was certainly not Darwin's concern. Having said this, something had to lead to Hitler and obviously the nineteenth century bears much of the guilt. So if you feel a slight sense of unease, I would not say that you are without justification, remembering of course that Darwin himself was not the only (some might say not the chief) foundation for Social Darwinism. (Richards 2008 discusses these issues in detail.)

A more positive Darwinian approach than that of the Nazis to the problems of the age, especially the appalling problems of the 1930s, was that of Julian Huxley, he of arms races fame. He was ever an evolutionist who thought that his science yielded significant ethical prescriptions. Not surprisingly, he thought that (particularly at the societal level) we should be promoting the virtues and benefits of science and technology. Responding to the Great Depression, Huxley is revealed as a great enthusiast for the public works funded by Franklin Roosevelt's New Deal. Although stepping somewhat warily because he did not want to be seen as endorsing the war preparations of the National Socialists – the building of the *Autobahn* for example – Huxley was fairly unrestrained in his encomia for the Tennessee Valley Authority, that project bringing electricity to large parts of the American South.

> All claims that the State has an intrinsically higher value than the individual are false. They turn out, on closer scrutiny, to be rationalizations or myths aimed at securing greater power or privilege for a limited group which controls the machinery of the State.
>
> On the other hand the individual is meaningless in isolation, and the possibilities of development and self-realization open to him are conditioned and limited by the nature of the social organization. The individual thus has duties and responsibilities as well as rights and privileges, or if you prefer it, finds certain outlets and satisfactions (such

as devotion to a cause, or participation in a joint enterprise) only in relation to the type of society in which he lives. (J. Huxley 1943, 138–39)

After the Second World War, Huxley became the first director-general of UNESCO, and indeed it was he who insisted that this new United Nations organization go beyond just concerns with education and culture, and include a concern with science also.

Finally let it be mentioned that today we find those still wanting to link evolution and ethics in a very traditional fashion. Edward O. Wilson has always felt this way, and recently this has been coming out in an understandable concern about the environment, specifically about biodiversity (E. Wilson 1984, 1992, 2006). He worries a great deal about the ways in which modern society is destroying the natural habitat and how with this comes the subsequent decline of natural resources and species diversity. Wilson sees humans as having evolved in symbiotic relationship with nature. Apart from the utilitarian factors – how for instance unknown, exotic species might produce substances of great social and medical benefit – Wilson believes that humans need the growing, living world in an almost aesthetic way. An environment of plastic would kill, literally as well as metaphorically. In a recent book, *The Future of Life*, Wilson declares: "a sense of genetic unity, kinship, and deep history are among the values that bond us to the living environment. They are survival mechanisms for us and our species. To conserve biological diversity is an investment in immortality" (E. Wilson 2002, 133).

Progress again

Now swing round and ask about foundations. Why did (or do) people like Spencer and Kropotkin and Jack London and von Bernhardi and Julian Huxley and Wilson feel so strongly that they could and must promote an evolutionarily based normative ethics? The answer – and by this stage you will hardly be surprised – is that they did it in the name of progress. Every one of these people was an ardent biological progressionist, as were their fellow-travelers (Ruse 2009b). We have seen this in detail in Spencer and Huxley and Wilson, and the same is true of the others. They saw the evolutionary process as one that had direction, leading up to humans. They also saw the evolutionary process as one gaining in value as it rose upward, and that humans represent the peak of excellence, biologically and in all other respects. Hence, they saw it as our moral duty to cherish and preserve humans, at minimum to keep us at the level we are now, and perhaps even to improve things for us in the future.

People like London and von Bernhardi thought of themselves as realists rather than blood-lust sadists. The struggle for existence is going to grind on, whatever you do. In the words of Spencer's American disciple, social scientist William Graham Sumner: "Man is born under the necessity of sustaining the existence he has received by an onerous struggle against nature, both to win what is essential to his life and to ward off what is prejudicial to it. He is born under a burden and a necessity. Nature holds what is essential to him, but she offers nothing gratuitously. He may win for his use what she holds, if he can" (Sumner 1914, 17). Better therefore to channel and direct it if one is able, and if not then at least not to stand in its way and prevent the better from succeeding and continuing. It is clear that

> [T]he possession of social power in any society or in any generation, produces social movement, with expansion, reiterated new achievement, social hope and enthusiasm, with all that we call progress; and that this movement is so directed that degradation is behind it. The problem is not to account for degradation, because if we relax our efforts we shall fall back into it. The problem is how to maintain the effort and develop the power so as to keep up the movement away from it. (150)

Philosophers have a standard reply against all of this. It was mentioned at the end of the chapter on progress. It comes from David Hume. One cannot derive the way that one would like the world to be, the way that one thinks the world *ought* to be, from the way that things are, from what the world *is*.

> In every system of morality, which I have hitherto met with, I have always remark'd, that the author proceeds for some time in the ordinary way of reasoning, and establishes the being of a God, or makes observations concerning human affairs; when of a sudden I am surpriz'd to find, that instead of the usual copulations of propositions, is, and is not, I meet with no proposition that is not connected with an ought, or an ought not. This change is imperceptible; but is, however, of the last consequence. For as this ought, or ought not, expresses some new relation or affirmation, 'tis

necessary that it shou'd be observ'd and explain'd; and at the same time
that a reason should be given, for what seems altogether inconceivable,
how this new relation can be a deduction from others, which are entirely
different from it. But as authors do not commonly use this precaution, I
shall presume to recommend it to the readers; and am persuaded, that
this small attention wou'd subvert all the vulgar systems of morality, and
let us see, that the distinction of vice and virtue is not founded merely on
the relations of objects, nor is perceiv'd by reason. (Hume [1739/40] 1978,
III, 2, i)

The early twentieth-century philosopher G. E. Moore (1903) spoke of making the move from "is" to "ought" as committing what he called the "naturalistic fallacy," trying illicitly to link natural facts with non-natural
facts, the latter including morality. In the case of traditional evolutionary
ethics, one is going from the way that evolution operates – the fittest survive and reproduce – to what one thinks should be the case – it is best that
the fittest survive and reproduce.

My experience is that most traditional evolutionary ethicists are profoundly unworried by this criticism. They argue that perhaps this is a mistaken move in most cases, but that in the evolutionary case it is legitimate.
The fact that the language changes (from "is" to "ought") is regarded as
no big deal. After all, this is always happening in science. The move from
talking about molecules buzzing around in a container to talk about temperature and pressure of a gas is a classic example. One is "reducing" gas
talk to molecule talk and likewise one is reducing ethics talk to evolution
talk. Evolution is progressive, humans won, our moral obligations are to
keep this up and to prevent decline.

A much better critical response, obviously, is to go after the notion of
progress and most particularly to go after the notion that tells us, if there
be biological progress, we have learned something about what is morally
worthwhile. Even if it does not stop them – I very much doubt that anything is going to stop an enthusiastic traditional evolutionary ethicist
(said with the weary knowledge of one who once co-authored an essay on
ethics with Edward O. Wilson) – it shows with devastating accuracy the
problems with the whole approach. Thomas Henry Huxley, for all that he
probably endorsed some version of the approach at one point, was one who
held the ideas up to the light, showing the threadbare nature of the fabric.
In a final, great essay on evolution and ethics, he pinpointed the fallacy

that lies behind the whole approach. "It is the notion that because, on the whole, animals and plants have advanced in perfection of organization by means of the struggle for existence and the consequent 'survival of the fittest'; therefore men in society, men as ethical beings, must look to the same process to help them towards perfection" (T. Huxley [1893] 2009, 80). He then went on to argue that although the aggressive features of the lion or the tiger may have served us well in the struggle for existence, it is precisely these that need to be quelled in the striving for civilization and moral worth.

> Men in society are undoubtedly subject to the cosmic process. As among other animals, multiplication goes on without cessation, and involves severe competition for the means of support. The struggle for existence tends to eliminate those less fitted to adapt themselves to the circumstances of their existence. The strongest, the most self-assertive, tend to tread down the weaker. But the influence of the cosmic process on the evolution of society is the greater the more rudimentary its civilization. Social progress means a checking of the cosmic process at every step and the substitution for it of another, which may be called the ethical process; the end of which is not the survival of those who may happen to be the fittest, in respect of the whole of the conditions which obtain, but of those who are ethically the best. (81)

The late-nineteenth-century Cambridge philosopher (and teacher of Moore) Henry Sidgwick was another who pinpointed the problems. Neither success in the struggle (the Darwinian version) nor complexity or heterogeneity (the Spencerian version) seems to have much to do with morality as such.

> It is more necessary to argue that the theory of Evolution, thus widely understood, has little or no bearing upon ethics. It is commonly supposed that it is of great importance in ethical controversy to prove that the Moral Faculty is derivative and not original: and there can be little doubt that this conclusion follows from the theory which we are now considering. For when we trace back in thought the series of organisms of which man is the final result, we must – at some point or other, it matters not where – come to a living being (whether called Man or not) devoid of moral consciousness; and between this point and that at which the moral faculty clearly presents itself, we must suppose a transition-period in which the distinctly moral consciousness is gradually being derived and

developed out of more primitive feelings and cognitions. All this seems necessarily involved in the acceptance of Evolution in any form; but when it is all admitted, I cannot see that any argument is gained for or against any particular ethical doctrine. For all the competing and conflicting moral principles that men have anywhere assumed must be equally derivative: and the mere recognition of their derivativeness, apart from any particular theory as to the *modus derivandi*, cannot supply us with any criterion for distinguishing true moral principles from false. (Sidgwick 1876, 54)

He added, making a point that will be of interest in relation to something discussed shortly, that one might rather think that because ethics is something exercised by beings who have evolved, one might more readily think that this shows that there can be no objectively grounded moral principles ("ethical skepticism"), but he added he did not see this conclusion as following at all. If well taken, it undercuts all knowledge.

It is perhaps more natural to think that this recognition must influence the mind in the direction of general moral scepticism. But surely there can be no reason why we should single out for distrust the enunciations of the moral faculty, merely because it is the outcome of a long process of development. Such a line of argument would leave us no faculty stable and trustworthy: and would therefore end by destroying its own premisses. (54)

The evolution of altruism

Taken as a whole, traditional evolutionary ethics is nowhere like as bad as its reputation. It would indeed be hard to live up (or down) to that reputation. However, although good things (as well as bad things) have been claimed in the name of evolution, metaethically it just doesn't work. Can we try another approach, perhaps along the lines of the more literal approach in evolutionary epistemology? Sidgwick (like most philosophers even today, I suspect) does not think that knowing about the origins of morality, or the moral organ as one might call it, tells us much, if anything, about the philosopher's questions on the topic. Is this perhaps too quick a decision? Nietzsche would have thought it was. "We need a *critique* of moral values, *the value of these values themselves must first be called in question* – and for that there is needed a knowledge of the conditions and circumstances in which they grew, under which they evolved and changed" (Nietzsche 1887, Prologue, 3). There are those today who agree with this sentiment. So let us start first with the science and then move on to the philosophy. Is it reasonable, in the light of Darwinian theory, to think that an ethical sense or ability might have evolved?

By now I think no one would deny an affirmative answer. Altruism in the biological sense is very well documented, and humans as completely social animals are very much in need of such altruism. If they had not developed it, they would be in deep trouble. But why, still speaking biologically, have we gone the extra step into literal altruism? Why are we not just like the ants, which seem to be programmed to do what they do, without any need of moral training or anything else of that nature? In one sense, presumably, the answer is "because we can." But that can hardly be the end of the matter. Surely, without dillying and dallying about these things, a lot of the time we would be far better off if we got on with things without thinking? A more satisfactory answer comes in another sense, where we appreciate the fact that humans have taken a reproductive route very different from most other organisms, especially the ants. For better or for worse, we simply cannot have many offspring. The physical needs of our children preclude that. Unlike the ants, where the queen can have literally hundreds of thousands of offspring, we humans are limited to about ten, give or take – and recognizing that the figures for men might be very different from those for women. What this means is that, if circumstances change, either the environment or fellow species members, we need to respond in a way that is not demanded of the ants. A queen ant can afford to lose a few thousand offspring in a shower of rain, washing away the pheromones that lead the nest members home. We cannot afford to lose even one child if it rains when they are away from home. We need the flexibility to rethink and reassess if things go wrong. That is what innate dispositions are all about. And in no place is this ability more crucial than when dealing with fellow humans and deciding what we should do in social situations, and what they should do also, and what we think they should do, and so forth.

Morality on this biological scenario, therefore, is an aid to decision-making in social groups. Should I help? Should others help me? Should I expect help? And so forth. In a way, it is all a bit like language, another social facilitator. Moral ability is innate, but we have to learn how to use it, and once acquired it is difficult to change it. It is the same as learning English or French. Perhaps you can push the analogy a bit further, suggesting that there is some kind of deep universal moral grammar and that growing up in a particular society you learn to apply it in somewhat different ways. People in an American small town in the Midwest share the deep structure with folk growing up under the Taliban in Afghanistan – reciprocation, and so forth – but how we learn to apply morality, say with respect to the role of women, is clearly going to be very different.

A major criticism which is often brought against any kind of biological approach to humankind is that this locks us into a deterministic view of human nature. It is argued that we deny free will and see humans as if they were marionettes, dangling on strings controlled by the DNA. However, note that although there are deterministic aspects to this picture, it is far from one of crude "genetic determinism." We are determined by biology to have the dispositions that we have and by our culture in the ways that these are expressed. All of us, perhaps, have the feeling that we ought to care for our own children, but whereas I have the feeling that you ought to be tolerant about the religious views of others, you perhaps have the feeling that one ought to persecute infidels. These are sentiments laid upon us. But no one, other than perhaps French existentialists at their most extreme (and unconvincing), has ever argued otherwise. Where we do have choice is whether we are going to obey the moral dictates. Sometimes, perhaps usually, we do the moral thing, but not always. In this sense, the only worthwhile sense, we are free in a way that the ants, which truly are genetically determined, are not free (Ruse 1987b).

What should we do?

Turn now to the philosophical questions. First, what about normative morality. Note that the Darwinian is going to be asking about what we think we should do as opposed to asking what we really should do. These are not obviously one and the same, so keep this point in mind. The answer is going to be couched in terms something along the lines of common-sense morality. That is the whole point of what *we think* we should do. In other words, morality is going to be something along the lines of the Love Commandment. (Treat your neighbor as yourself, or do as you would be done by.) Note that for all that philosophers delight in finding odd examples that seem to separate different moral systems – if you are in a concentration camp, is it moral to bribe a guard with chocolate in order to escape the gas chamber even though by doing so you are corrupting another human being? – by and large the secular moralities agree with each other and with religious systems on day-to-day things. Everyone agrees that you ought to be kind to small children and respect your mum and dad and not mark up library books. Where differences come is when people differ on factual matters. Everyone agrees that killing another human being just because it makes things easier is wrong. The abortion debate is over whether a fetus is another human being.

Of course, one problem is always that of sorting out what we say we think from what we actually think. Most of us would say that you ought not to cheat on your taxes, but whether we stick to that is another matter – or more precisely, it is another matter whether we really feel that diddling the government out of a few bucks is really such a bad thing. Evolutionary biology predicts that we are going to take reciprocation pretty seriously, and in fact that does seem to be the case. We want people to be fair. It is not so much paying taxes that we object to, as the feeling that people much richer than we get out of paying their share or any tax at all. This in fact ties in with much that moral philosophers (in the Anglo-American tradition) have been arguing for the past half century. The Harvard philosopher John Rawls (1971) argued that we ought to be just and to be just was to be fair. How would this cash out in society? Rawls invited us to pretend that we did not know what position or role we were going to have in society. We would not know if we would be male and rich and healthy or female and poor and unhealthy. We were behind a "veil of ignorance." If we did know, then we would want our own group to be favored. But because we do not know, we would want society to do the best for us, whatever our position. In other words, we would want a society that would look after the suffering, even at the expense of the fortunate. Not that everyone would necessarily get the same. You might pay your firemen more than your garbage collectors, because theirs is the more hazardous occupation and you want the best-qualified people to take it. But everyone would see that the additional pay benefits us all – them directly and us indirectly by having the best firemen possible.

Explicitly Rawls linked this to Darwinian evolutionary theory:

> In arguing for the greater stability of the principles of justice I have
> assumed that certain psychological laws are true, or approximately
> so. I shall not pursue the question of stability beyond this point. We
> may note however that one might ask how it is that human beings
> have acquired a nature described by these psychological principles.
> The theory of evolution would suggest that it is the outcome of natural
> selection; the capacity for a sense of justice and the moral feelings is an
> adaptation of mankind to its place in nature. As ethologists maintain,
> the behavior patterns of a species, and the psychological mechanisms of
> their acquisition, are just as much its characteristics as are the distinctive
> features of its bodily structures; and these patterns of behavior have an
> evolution exactly as organs and bones do. It seems clear that for members
> of species which lives in stable social groups, the ability to comply with
> fair cooperative arrangements and to develop the sentiments necessary to
> support them is highly advantageous, especially when individuals have a
> long life and are dependent on one another. These conditions guarantee
> innumerable occasions when mutual justice consistently adhered to is
> beneficial to all parties. (Rawls 1971, 502–503)

Agree that perhaps (possibly, certainly?) an evolutionary approach to normative morality can yield a fairly conventional set of rules. Is there anything in the approach that might speak to its virtues, convince us that
it is worth taking seriously? Can it solve problems in more satisfactorily
than other approaches? In recent years, moral philosophers have rather
obsessed over a number of paradoxes, the most famous of which is probably the so-called "trolley problem." Suppose you are down a mine, and
a runaway coal truck is coming down the line. Five unaware people are
standing on the main line. You are by a switch and you could easily pull it,
diverting the truck to a side-line. However, on the side-line stands another
unaware person. Would you pull? Now suppose the same situation, but
there is no switch and no side-line. There is a big person standing next to
you, and you could push this person on the line and stop the truck. (You
are just a runt, so it is no good throwing yourself on the line.) Would you
push? Most people say they would pull but not push. Yet this seems incon-
sistent, for in both cases you are sacrificing one to save five.

Evolutionists argue that here they can step in positively and help (Singer
2005). The situation is very much like the paradoxes revealed by the
Wason test, where likewise we had different actions or beliefs, although

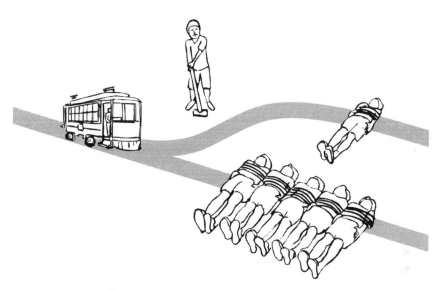

Figure 6.1 The trolley problem. Why are people willing to pull a switch and save five at the expense of one, but unwilling to push someone on the line to save five at the expense of one?

the situation was formally identical. The fact is that our past leads us to care for those in our immediate vicinity. Those of our would-be ancestors who looked to the immediate interests of their near neighbors tended to survive and reproduce – whatever rationality might predict about the future – and those who did not, did not. The trolley case is artificial, to say the least. Normally just helping our neighbor is not something that kills five other people. Conversely, on the other supposition, just pulling a switch is not something that triggers evolutionary barriers and emotions, and so we can contemplate doing this. Backing this kind of thinking is the discovery that we use different parts of our brain when we are thinking about the two cases (Greene *et al.* 2001; Greene and Haidt 2002). One part, the emotion-producing part, deals with the push case. Another part, the reason-producing part, deals with the pull case.

The same strength of the evolutionary approach emerges when we are dealing with what I have called the "*Bleak House*" case (Ruse 2009b). To whom do we owe the greater obligation? Our children or a neighbor? A neighbor or a stranger? Some people have argued that we have the same obligation to all. But one wonders really if that is so. In his great novel,

Dickens is scathing about those who neglect family over others and who neglect the poor of London, the poor of our own society – Jo the crossing-sweeper – over the needs of those in distant lands. Hume, even if not an evolutionist, put his finger on the issue. Morality follows relationships. "A man naturally loves his children better than his nephews, his nephews better than his cousins, his cousins better than strangers, where everything else is equal. Hence arise our common measures of duty, in preferring the one to the other. Our sense of duty always follows the common and natural course of our passions" (Hume [1739/40] 1978, III, 2, i). Charity begins at home and Darwinian evolutionary theory explains why we think this is so.

Ethical skepticism

What about foundations? Let us agree that evolutionary biology can tell us much about what we think we should do. But this is not to tell us what we really should do, and for this we need to get to foundations. What is the metaethical justification that the evolutionary biologist is offering? It cannot be progress, because that simply will not work. But what else could it be, if you are offering a naturalistic foundation? There is one possibility, namely that there is no foundation at all! Perhaps what we believe to be the case is all that there is to it. There is no reality, no objective truth, beneath what we think is the truth.

Philosopher Jeffrie Murphy has argued precisely this.

> The sociobiologist may well agree with the point ... that value judgments are properly defended in terms of other value judgments until we reach some that are fundamental. All of this, in a sense, is the giving of *reasons*. However, suppose we seriously raise the question of why these fundamental judgments are regarded as fundamental. There may be only a *causal* explanation for this! We reject simplistic utilitarianism because it entails consequences that are morally counterintuitive, or we embrace a Rawlsian theory of justice because it systematizes (places in "reflective equilibrium") our pretheoretical convictions. But what is the status of those intuitions or convictions? Perhaps there is nothing more to be said for them than that they involve deep preferences (or patterns of preference) built into our biological nature. If this is so, then at a very fundamental point the reasons/causes (and the belief that we ought/really

ought) distinction breaks down, or the one transforms into the other. (J. Murphy 1982, 112n)

Even if true, this is all a bit quick. The argument is for what philosophers call "moral non-realism" or sometimes "ethical skepticism." In other words, the claim is that moral claims do not refer to anything and so strictly speaking are false. If I say "rape is wrong" then I am not talking about something "wrong" and so strictly speaking what I am saying is false. (This is contrasted with "non-cognitivism," the logical positivists' ethical position, which claims that moral claims are meaningless and hence only express emotions.) Of course, this is not to deny that within the system (rather like the epistemologist's coherence position) one can make true and false statements, just as in baseball one can make true and false statements. In baseball, three strikes and you are out. In normative morality, rape is wrong. The point is that in neither case are you talking about anything out there. Note, however, that the moral non-realist is making a stronger statement than just the epistemologist's coherence case. The point is that, however you regard the nature of material objects and the truth status of scientific claims (coherence or correspondence), the claims of ethics are weaker in a sense. Not in the sense that we take them less seriously or more relativistically – you can still be as hard-line on rape as on gravity – but that whereas claims about the physical world refer (if only in an internal realist sense), claims about the moral world do not.

How can you make this case? In one sense, I am not sure that you can. What would it mean to be a moral realist? You are referring to moral facts, whatever they are. And what they are is certainly, as Moore (1903) pointed out, non-natural. They have to be something existing in their own right or the Will of God or some such thing. Now whether any of this even makes sense is by the by – some philosophers have thought not, some philosophers think that it does – the point is that working in the natural world with evolutionary theory, you are simply not talking about the non-natural directly. So if someone insists that, even though morality as we know it at the normative level can be explained absolutely and completely in terms of evolutionary biology, nevertheless they believe that God stands behind the whole system making it work, qua naturalist I don't think you can put a finger on them (Ruse 2010).

However, I am not sure that this is quite the end of the discussion. For a start, if the evolutionary account works, then the objective reference is not needed. Normative morality is as it is because of the biology, and if you don't want to go beyond the natural you do not have to. I suspect that the Christian, let us say, is not going to find this terribly worrisome. The Christian does want to go beyond the natural, and that is all there is to it. However, my suspicion is that someone who is completely secular might find this worrisome. Non-natural properties seem now to be redundant, so what are they doing and why believe in them? Think about the analogy with mathematics for a moment. Many with strong naturalistic convictions nevertheless think that there might be something to mathematical Platonism, believing that mathematical objects do exist in some non-natural realm. Why? Because they simply cannot think of a way that mathematical truths can be explained naturally. Sure, we get basic arithmetic in the ways suggested in the last chapter, but what about (what you will by now rightly infer I consider one of the most beautiful things ever encountered by humankind) the Euler identity, $e^{\pi i} + 1 = 0$? Can this be fully explained (or explained away) by our biology? If it cannot, even though you are now opening up a can of worms about how the non-natural can nevertheless affect or control the natural, at least you have a good reason to think in terms of non-natural entities. But assuming that the Darwinian explanation does work, this argument is not open in the case of ethics.

Actually the situation is worse even than this. If you take seriously the non-directedness of evolution, then there is really no reason to think that the ethical system possessed by humans (for the sake of argument, let us agree that there is at least some shared "deep structure" to normative ethics, as there is in linguistics) is necessarily going to be the only one or that the one we have is the right one (assuming that there is some independent objective morality). Kant ([1785] 1959) did not think normative morality referred to anything as such (although perhaps God is involved somewhere) but he did think that morality is necessary (in the synthetic a priori sense), because the morality we have is a necessary condition for rational beings to work together. If no one thought lying is a transgression, then society would not work. However, although we might perhaps agree that there are certain formal rules that must be obeyed in successful societal systems (the sorts of things that game theorists take seriously), filling in these formal rules could take us in very different directions. (Which

Kant himself actually realized.) We need to interact and work together. One way is through shared trust – I will help you and you will help me, and the reason why we do this is because we think we should help each other (the Golden Rule).

Perhaps another way is through shared mistrust. I dislike you intensely; more than this my dislike is fueled by a moral sense, so I don't weaken. However, I know you feel the same way about me and so it is in our interests to cooperate because otherwise we are constantly at each other's throats. I have on occasion called this the "John Foster Dulles" system of morality after President Eisenhower's Secretary of State in the Cold War (Ruse 1986). He hated the Russians with a moral passion. They felt the same way about him. And so there was arms control. The point is that evolution might have taken us this way, so the real non-natural morality ("love your neighbor") is unknown and unused. That does seem to be a reductio of what we normally mean by moral facts, natural or non-natural. Perhaps the Christian will argue that God would never let this happen, but that raises other issues about whether you want God actively involved in guiding evolution.

Objectification

One final point. Suppose the moral non-realist is right. Suppose there is no objective reality. Why do most people, including most philosophers, find this so hard to accept? To the Darwinian, the answer is obvious. If we did not believe that morality was objective, that it refers to real facts, then we would soon quit obeying it (Mackie 1977, 1979). Why should I be good when there is no reason? At least, no reason in my self-interest as against simply looking after Number One? So, to use an ugly word introduced by the late John Mackie, we "objectify" normative morality, thinking it does have a foundation, even though it does not. This is a case where biology is deceiving us for our own good, because Kant was right: if we stop being moral society breaks down and then we all lose.

So now that I have told you, why don't you go away and be immoral, pretending of course that you are being moral? Because we are not psychopaths. Our psychology would make us very uncomfortable were we to do this. Dostoyevsky showed this in *Crime and Punishment*. The police chief knows that the student Raskolnikov has murdered the two old women, because he thought he could transcend his moral nature. But in the end, the student confesses because he cannot go against himself. As the Freudian argues that you are denying the truth of the Freudian analysis because of your own problems, so the Darwinian argues that you assert the objective truth of morality because of your own nature, one that Darwinism has brought about! You can't win? Oh yes you can, but only if you stay within the system!

7   Sex, orientation, and race

In this chapter I want to look at three questions on human nature about which Darwinians have had things to say – things to say that others have not always received entirely favorably. Male–female differences; sexual orientation; and race. Let us take these in turn.

Some background

By "sex" one generally means the biology. Males have beards and penises; females have breasts and vaginas. Males have prostates; women have wombs. Males have an X and a Y chromosome; females have two X chromosomes. (Birds reverse this pattern.) In sexuality, genetic information is being combined in one organism from two earlier organisms. By "gender" one generally means the psychology. Males identify with the biology of males; females with the biology of females. Transsexuals have sex and gender crossed. Most people are obviously not transsexuals and so generally when I use the word "sex" I am including gender. Hermaphrodites have both sexes and need another to reproduce. Asexual organisms do it themselves, without need for others. (Some organisms vary between sexuality and asexuality, according to the season or food supplies or whatever. Overall, these are generally considered sexual, because at some point genetic information is being pooled.) Sexual orientation is about the type of person (sex, which usually coincides with gender) with whom you want to have sex. There is no assumption that someone who wants to have sex with someone of their own sex has any issues about gender. A male homosexual can (and almost always does) feel perfectly comfortable about being a man, and similarly for women.

We joke about Freud and sex. That is nothing to Darwin and sex. As you will realize by now, evolutionary biologists – Darwinians in particular – are absolutely obsessed with sex and have had much to say on the subject. What would you expect from a theory that puts reproduction right up at the top of things needed for evolutionary change? To a certain extent, the obsession with sex reveals our own human interests. If you look at a modern tree of life, most organisms are not in the sex business. It is our branch which makes a big deal of it, which at once raises the question: why sex? Why don't all organisms reproduce asexually? This question brings us right up against one of the major issues of recent evolutionary biology (Maynard Smith 1978). Having sex requires effort. The actual act may be all fun and games, but the work leading up to it is not trivial. And then after sex, if and when the female has been impregnated, what's in it for her? In many cases, the male is conspicuous only by his absence. Why should she waste all of that effort if 50 percent of the genes being transmitted are not hers? And yet there does seem to be a reason for sex. In species where you get both sexual and asexual forms, by and large the asexual groups don't last (and have to be created again from sexual organisms).

The most obvious reason for sex, and there is surely some truth in this, is that sex exists because in that way useful new variations (mutations) can be gathered together quickly in one organism. In asexual organisms, if you get good mutation A, you might have to hang around a long time before you also get good mutation B in the descendants. But with sexual organisms, if one organism gets A and another gets B, then in a generation or two you can get an organism with A and B. The trouble with this kind of explanation, however, is that it relies on group selection operating. Combining the genes is great for the population, but what's in it for single members of that population? It is no big benefit to an individual female to raise kids who are only half related to her. Why is she not eliminated quickly by females who raise kids completely related to them? There have been a number of responses to this problem, and I think it fair to say that not one of them has commanded universal approval, to the extent that a lot of evolutionary biologists probably doubt that there is going to be one simple solution. Sex is likely to be one of those messy issues best dealt with on a case-by-case basis.

For a start, it seems agreed that you had better split up the problem of why sex starts at all and the problem of why sex is maintained once it has started. Suppose that there is something in the combining genes answer, at least for getting things going before selfishness (in the biological sense) can kick in. That would give you an answer to why sex starts, even if now you have to answer why sex stays on and on. George C. Williams (1975), one of the major players in recent discussions on this topic, took a somewhat gloomy view of the situation. Whatever the early benefits of sex, he thought that today's higher organisms like mammals are basically stuck with it, even though it may well be maladaptive. Humans would be better off reproducing asexually, but because of our physiology and anatomy, it is no longer possible. Most do not go that far, and I should say that Williams was an exemplary husband and father! Generally, at least for maintenance, opinion divides. Virtually no one wants to say that group selection is absolutely impossible, and perhaps this is one of those cases where it does operate. If the immediate benefits are so great and if populations combine and split up on a fairly regular basis, then group selection could work. Perhaps, however, individual selection plays a major role. William Hamilton, in one of those leaps of imagination that made him the leading evolutionary biologist of the past fifty years, suggested that perhaps we have an arms race, with slower-evolving organisms trying to stay ahead of faster-evolving parasites (Hamilton *et al.* 1990). Parasites generally can and do change more rapidly than their hosts – more on this in the next chapter, but think of the speed with which venereal diseases have responded to penicillin and other post-World War Two drugs. Hamilton argued that sex is a means to shake up the genotype and thus present the parasite with a constantly moving target. It is not so much new mutations that count, but recombining in new arrays the already existing genes. He went on to back up this audacious hypothesis with related evidence, for instance about the ways in which sexual attraction can exhibit physical fitness – bright-red displays showing healthy blood was one example he used (Hamilton and Zuk 1982).

Darwin on sex

Whatever the reasons, members of the branch of life to which we humans belong have sex. Even if overall having sex (being sexual) is a bad thing, we are stuck with it; so we expect natural selection to make the best of a bad job. Or sexual selection, rather. It is this mechanism that kicks in where sexuality is concerned, sex differences especially. We saw in an earlier chapter that sexual selection was not just some ad hoc add-on, but something that was part of Darwin's theory from the first, undoubtedly a corollary stemming from Darwin's use of the artificial–natural selection analogy to get to his main mechanism in the first place. However, it was not until the *Descent of Man* that Darwin made much of the secondary mechanism, and this in response to Alfred Russel Wallace going soft on naturalistic explanations of the origins of humankind. I should say that, after the *Descent*, sexual selection did not bloom particularly – natural selection had enough trouble finding its way – and even those who accepted selection generally tended to roll sexual selection into natural selection considered as the cause of change. It was not until the 1960s and the rise of the individual selection perspective, combined with ever-more detailed and lengthy studies of animal behavior, that sexual selection started to rise in general importance in the eyes of evolutionary biologists (Campbell 1972).

There were a number of outstanding empirical studies. One was that of the British biologist Geoffrey Parker on the lowly dung fly, showing how the insects vary their behaviors to attract mates (Parker 1978; Ruse 1999). Another, also by a British biologist, Tim Clutton-Brock, was on red deer on an island off the coast of Scotland (Clutton-Brock *et al.* 1982). He demonstrated in great detail how the stags fight each other, competing to take over the harems, and how this pays off in reproductive success and failure. It should be noted that the very idea of sexual selection has its critics, notably the Stanford evolutionary ecologist Joan Roughgarden (2009), who thinks that the whole notion is flawed, and sexist to boot. However, to adapt Jane Austen, this is not a truth universally acknowledged. A major worry of Roughgarden's is that sexual selection reifies sexual differences (that is, accepts that these are objective aspects of the world), and she thinks this is unfair to transsexuals, who obviously have a more fluid view of these things. But surely accepting that there are some borderline cases does not deny the reality of most cases. Whether or not Pluto is a planet, Earth and Mars and the others are planets. Jack Kennedy was really male and Marilyn Monroe was really female.

We will be coming shortly to the question of sexual orientation, but relatedly Roughgarden thinks that sexual selection is inadequate here. There is far more same-sex activity in the animal world than can be explained by generally accepted mechanisms, and so something else is needed. This something else requires cooperation between organisms, as they work together for the common good.

> Many animal behaviors involving physical intimacy, such as grooming, traveling, and sleeping in close proximity, making reciprocal interlocking vocalizations, and same-sex and between-sex sexuality could all promote coordinated action. Further, we hypothesize that a sense of friendship resides in animal bonding, a joy or synergy in the spirit of cooperation that allows animals to sense and experience the product, not merely the sum, of their individual well-beings. (Roughgarden *et al.* 2006, 967)

Needless to say, this has gone over with the average Darwinian like the proverbial lead balloon (Clutton-Brock 2007). Apart from being grossly, some would say grotesquely, group selectionist, talk of "joy or synergy in the spirit of cooperation" gives scientists the creeps. If ever there was a case of the Californian flower-child philosophy being read into the animal world, this is it.

Let us pick up again on sexual selection, which it will be remembered was divided into two kinds by Darwin (although he did stress that this was not a division without exceptions). There is selection through male combat (as with the deer's antlers) and selection through female choice (as with the peacock's tail). This leads to physical and behavioral differences between males and females, although as far as Darwin was concerned, there is a certain asymmetry here.

> Throughout the animal kingdom, when the sexes differ from each other in external appearance, it is the male which, with rare exceptions, has been chiefly modified; for the female still remains more like the young of her own species, and more like the other members of the same group. The cause of this seems to lie in the males of almost all animals having stronger passions than the females. Hence it is the males that fight together and sedulously display their charms before the females; and those which are victorious transmit their superiority to their male offspring. (C. Darwin 1871, I, 272)

He continues: "The female, on the other hand, with the rarest exception, is less eager than the male. As the illustrious Hunter long ago observed, she generally 'requires to be courted'; she is coy, and may often be seen endeavouring for a long time to escape from the male." Nevertheless, don't think that females do nothing: "the female, though comparatively passive, generally exerts some choice and accepts one male in preference to others. Or she may accept, as appearances would sometimes lead us to believe, not the male which is the most attractive to her, but the one which is the least distasteful. The exertion of some choice on the part of the female seems almost as general a law as the eagerness of the male" (I, 271–73).

All of this is then applied straight to humans.

> Man is more courageous, pugnacious, and energetic than woman, and has a more inventive genius. His brain is absolutely larger, but whether relatively to the larger size of his body, in comparison with that of woman, has not, I believe, been fully ascertained. In woman the face is rounder; the jaws and the base of the skull smaller; the outlines of her body rounder, in parts more prominent; and her pelvis is broader than in man; but this latter character may perhaps be considered rather as a primary than a secondary sexual character. She comes to maturity at an earlier age than man. (C. Darwin 1871, II, 316–17)

There is quite a bit more in the same vein.

> The chief distinction in the intellectual powers of the two sexes is shewn by man attaining to a higher eminence, in whatever he takes up, than woman can attain – whether requiring deep thought, reason, or imagination, or merely the use of the senses and hands. If two lists were made of the most eminent men and women in poetry, painting, sculpture, music, – comprising composition and performance, history, science, and philosophy, with half-a-dozen names under each subject, the two lists would not bear comparison. (C. Darwin 1871, II, 327)

All of which is a matter of sexual selection at work, and of the bravest braves getting the pick of the crop, subject of course to the women also trying to work things to their own ends. Don't forget that having the Number One male sire your children has virtues right then and there. Not that things are always going to end up quite as we Europeans might expect. Remember those Hottentot women with the big bottoms.

Is Darwinism sexist?

You can imagine how feminists react to all of this. Admittedly it is in the later work that Darwin really lets it all hang out. "The *Descent* gives voice to Darwin's deeply rooted beliefs" (Erskine 1995, 100). But this, we learn, is only a cover for the theory of the *Origin*. "If his *Origin* statements appear neutral, it is only because patriarchy and the subordination of women were for him unchallenged assumptions" (Erskine 1995, 100). Hence:

> The *Origin*, seen in the wider context of Darwin's views on women, implies female subordination. The central focus on sexual reproduction, and the female's role as a vessel for the development of the next generation, meant that success in the role took on primary importance. Sexual selection forced males to become ever stronger and fitter, whilst making females progressively more passive. (101)

In short: "The *Origin* provided a mechanism for converting culturally entrenched ideas of female inferiority into permanent, biologically determined, sexual hierarchy" (118). Or more simply in the words of another critic: "Darwin's theory of evolution thus perpetuates the view of woman as less perfect than man" (Tuana 1993, 38).

Stirring stuff and charges easy to make about a man long dead, but is it true? There are two issues here. First, is sexual selection theory, and by implication the whole theory of the *Origin*, implicitly or explicitly sexist, in the sense of demeaning to women and basing judgments on values not found in the empirical world? Second, what about Darwin himself? The second question is easier to answer. There is no question but that Darwin reflected here the standards and values of the comfortable, mid-Victorian middle class to which he belonged. You have only to read Charles Dickens to see the parallels. Darwin was reading an awful lot in and then reading an awful lot out. He was quite explicit that somehow women should be regarded as less-developed men, and the fact that he extended (as he did) this courtesy to other animals does not alter the point. However, we should be careful with blanket condemnations, judgments from the present, for otherwise we end up criticizing every society that is not our own for failing to live up to our standards. More importantly, although Darwin was certainly not as far advanced (toward our thinking) as some – John Stuart Mill, for example – in many respects he was far more liberal than others; in, for example, his position on the American Civil War, which has already been mentioned. So one should not think that he was going out of his way to be reactionary. Consider: Thomas Henry Huxley, whom all today would think of a progressive thinker who did much to bring the world forward from the eighteenth to the twentieth century, did not think that women should be allowed to attend medical schools (Desmond 1994, 1997). And also, as even feminist critics have agreed in the case of Freud, Darwin is giving us a snapshot of the sexual relationships of his own day, as much as anything. It is true that like Freud, he is dressing it up in the garb of objective science, but we can reject that and still gain insights into human nature.

The general question of sexual selection theory is more difficult and more interesting. Whatever you decide, it does seem a bit of a stretch to tar the whole of Darwinian thinking simply because it has implications for thinking about male–female differences. Or put it this way, if in looking at evolutionary questions you are not going to talk about reproduction and the fact that it is females who have the offspring, then what are you going to talk about? Even if Joan Roughgarden's comments about homosexual cuddling were well taken, you need a bit more to get evolution. One could, I suppose, argue that the very inquiry into origins is tainted. No matter what the results, I suspect that any reviewer sent a proposal from the National Science Foundation on Jewish nose sizes and their links to avariciousness would feel that something was a bit morally queasy. But is the inquiry into origins really of this nature? Is the Nobel Prize winner Steven Weinberg (1977) a sexist for writing a book about the big bang? And if the big bang passes the purity test, why not plate tectonics? And if plate tectonics, why not the origin of the dinosaurs? And if not dinosaurs, why in principle should we not look at humans? Perhaps indeed, going on the offensive, one could argue that a scientific inquiry into human origins will replace the patriarchal account of origins given in Christianity – all of that stuff about Eve being an afterthought made out of Adam's rib to keep him occupied.

What about sexual selection? Here, history steps in to help the discussion. Alfred Russel Wallace (1905) was not only a socialist but also an ardent feminist. Generally, as noted earlier, he was not very keen on sexual selection, at least not on female choice (Wallace 1870). Perhaps this was part of his dislike of individual-selectionist arguments or perhaps he did not like the fact that female choice puts the glamor on the male. Either way, he argued that sexual dimorphism (in these sorts of cases) is more a matter of the females staying drab and camouflaged because it is they who have the young and who need to stay out of the way of predators. (Note that these sorts of cases often involve birds, where protection of the nest, something often in a vulnerable spot, is of prime importance.) However, when it came to humans, for Wallace things reversed themselves. Here he thought women take over, if not now then they in the future. Apparently, we will all rise upward because young women will choose as mates only the best and finest of young men, those worthy of love and respect.

> In such a reformed society the vicious man, the man of degraded taste or of feeble intellect, will have little chance of finding a wife, and his bad qualities will die out with himself. The most perfect and beautiful in body and mind will, on the other hand, be most sought and therefore be most likely to marry early, the less highly endowed later, and the least gifted in any way the latest of all, and this will be the case with both sexes. (Wallace 1900, II, 507)

Frankly, if this is based on personal observation within his own family, then the Wallace girls must have been as odd as their father. But this is not quite the point, which is rather that sexual selection as such lends itself to interpretations that are as far from sexist as it is possible to imagine. Nor is Wallace the only one to argue this way. The feminist biological anthropologist Sarah Hrdy (she who wrote on infanticide) has likewise used sexual selection to propose scenarios where it is the female in charge and the poor unsuspecting male who is left dangling (Hrdy 1981). Because human females do not come into heat, this means that their times of fertility are concealed and that hence males cannot easily tell if they are the fathers of the children of the women with whom they have had sexual intercourse. Women therefore can control men, keeping them around on a more or less permanent basis to provide child care. Moreover, females can to a great extent do the picking and choosing of those with whom they want to mate.

Much of this kind of argumentation seems to me about as well rooted in biology as Darwin's. However, there is a better-justified moral to be extracted. In an important sense, Darwinian biology tells you that there cannot be anything inherently sexist in sexual selection. If it favored males over females, then mothers would have only males. But very quickly, as of course is happening or is bound to happen very soon in societies like India and China that have been using modern technology to skew the ratio in favor of males, the premium would be on females. It is a seller's market. There is no point in having sons if there are no females with whom they can mate. So males and females have to have skills to balance things out. Parenthetically, there is an interesting point here about the difference between individual and group selection. In species where the males perform little or no parental care, from a group perspective you might be better off with just 10 percent males. That is enough for all of the gene shuffling and combining that you need. But from an individualist point of view, if there are only 10 percent males then a mother is better off having males. Soon the balance in numbers is restored. This is what we find in practice, subject to qualifications about skewed ratios coming from asymmetric effort being required to raise one sex over the other, and so forth.

You may not like the talk of aggressive males and coy females, but the point is that unless these strategies work, they will be dropped. And this leads straight into the reason why Darwinian biologists are going to persist in using such terms, or alternative language denoting the same things. It is the baby-bearing business again. The fact is that it is the females who have the babies. They are the limiting resource. And especially when you start to get up the scale to the vertebrates and to organisms like mammals and birds, this counts. Sophisticated organisms almost always require not just a relatively lengthy period of gestation, but also after-birth care. In a way, females are stuck with doing this, whether they want to or not. This does not mean that males cannot be brought into child care. Birds are a case in point, where there is obvious need to raise the offspring quickly and it is in the males' biological interests to get involved. And as always, there are total exceptions, as for instance among the sticklebacks, where it is the males who provide child care. (Expectedly, it is the males who are courted rather than vice versa.) But the overall point is that nature has skewed things. Remember the point made earlier (in chapter 3). Males can in theory have lots of offspring. Females can in theory have a limited number of offspring. However, going the other way, males may have lots of offspring, but they can end up with none. Females may have few offspring, but they are pretty much guaranteed to have these offspring. Hence, males are going to have features that help them to compete. Females are going to have features that help them to raise the best possible offspring, and that means being choosy.

Men and women compared

What does this mean for humans today? Two points are obviously pertinent. One, our past is going to matter. Two, modern society can and does change things. As far as the first point is concerned, answers are not going to be quite as easy as you might think, but they are there. Look at our closest relatives, starting with the gorilla. There you have a fairly straightforward example of sexual selection at work. Males have harems, males compete for the harems, females find it in their interests to go with the alpha males (if only for the genetic benefits for their own sons), and so we find significance sexual dimorphism. Males are much bigger and stronger than females. The chimpanzees on the other hand, especially the pigmy chimpanzees (the bonobos), have taken a somewhat different route. Males and females live much more socially together, sex is something shared through the group, females are much more obviously major players in the power struggle, and more. It doesn't mean that sexual selection is not at play – males are bigger than females – but it certainly is not the sort of society that Darwin thought held for humans. In the case of humans, no Darwinian biologist can or would deny that sexual selection has been at work. Males are about 20 percent bigger than females, and although gaps have narrowed, it is clear from sporting records and the like that males are stronger and faster, and so forth. Our testicles also tell a story (Harvey and May 1989). Chimpanzees are massively endowed, because given their social structure they need to produce lots of sperm. Gorillas are way down the list, not because they are not interested in sex, but because harem behavior means that you can make every drop count. We are somewhere right in the middle. In other words, you do look for a certain amount of aggression on the part of males and a certain amount of reticence combined with choosiness on that of females. And we would expect this to be revealed in moderate polygamous behavior – polygyny (one male, multiple females) rather than polyandry (multiple males, one female). This is precisely what social surveys of human societies reveal. You hardly ever get polyandry, for example, except in special circumstances – Tibetan farmers, Inuit hunters – where several men are needed to support the one family (van den Berghe 1979).

What does this imply for modern society? Is the Darwinian saying that monogamy is unnatural? Well, for a start, in Western societies we don't exactly have monogamy – serial monogamy is a better term (and perhaps serial polygamy is even better). In earlier societies there was a huge amount of remarriage after the early death of a spouse (childbirth being a regular factor in this), and today with easy divorce there is a much changing of mates, and by the very nature of things (in biological as well as social terms) it is more likely that the male will remarry and start a second family. What is more, a point made earlier in our discussions of culture, nothing in biology is written on stone. If (as happens in Western societies) women get empowered and freed from constant, unavoidable childbirth, you expect to find some shifts in behavior and overall patterns. What the Darwinian is likely to say is that one should be cautious about utopian proposals for complete sexual identity and – without necessarily buying into the whole Evolutionary Psychology program – recognize that the Pleistocene might still echo in (not dominate or forever determine) today's society. It might just be that women want to spend time with their young children in ways that men do not. The Darwinian might say that the moral course of action is not to pretend that this is not the case, or to try to brainwash people out of it, but to reorganize society so that these natural desires can be fulfilled for the best ends of all. This might well be put in the broader context of asking why it is, as the critics point out, that not just Darwin, but the whole of Western culture from Plato and Aristotle (usually highlighted as a particularly egregious sinner) onward has been so resolutely sexist, judged by today's standards. To what extent was constant, unavoidable childbirth a terrific, determining factor in male–female cultural differences, and to what extent is this something that once eliminated (through efficient birth control) leaves no trace whatsoever?

What about the $64,000 question of intelligence? Darwin obviously thought that females are lower down the scale than men, and one suspects that this was a view shared by many right through to at least the middle of the twentieth century. Boys would do mathematics, physics, and chemistry; girls would do geography, biology, and home economics. Now we know just how drastically mistaken a view that was. Today, at universities in the West, the undergraduate population is almost invariably (outside certain special cases, often specializing in technology) tilted about 60:40 in favor of females. And this goes over to mathematics and science. Writing about the USA, two (balanced) observers lay out the facts:

> Roughly half the population is female, and by most measures they are faring well academically. Consider that by age 25, over one-third of women have completed college (versus 29% of males); women outperform men in nearly all high school and college courses, including mathematics; women now comprise 48% of all college math majors; and women enter graduate and professional schools in numbers equal in most, but not all fields (currently women comprise 50% of MDs, 75% of veterinary medicine doctorates, 48% of life science PhDs, and 68% of psychology PhDs). (Ceci and Williams 2009, 5)

One thing you can be sure of is that this change is not a function of biology. Just take a look at the really remarkable statistics from the Cornell University Veterinary School with respect to admissions (Figure 7.1). There was no set of mutations around 1955 that led to the stupendous jump in females starting around 1975. No doubt in part there were social factors. Veterinary medicine changed from mainly rural large-animal practice to urban small-animal practice. These days, if a farmer cannot cure a sick cow with a home-administered shot of penicillin, he or she is often likely to ship it off to the knacker's yard at once. This was combined with the glamorizing of the profession by the book, film, and above all the television series *All Creatures Great and Small* – this was somewhat paradoxical, because the Yorkshire practice portrayed was mainly large-animal (although, note, set in an earlier time). But the emphasis above all was on the fact that being a vet is a human, caring way of life. However, the main reason for the jump was simply that the college stopped discriminating against women. Out went quotas and in came admission on abilities and qualifications. It was as simple as that.

None of this disproves Darwinian biology. Indeed, from what we have seen of today's thinking, if anything it confirms it. For a start, the evolutionary psychologists are stressing that reasoning is not some isolated, rather ethereal activity, but occurs very much in social situations. There is no reason to think that females would be any less endowed than males. (If selection is working to keep the sexes equal, the comparative sizes of the brains are probably not as significant as one might think. Look at how today's computers can pack a huge amount more into much less than those of even a couple of decades ago.) Think next of the chimpanzees, another very social species with males and females mixing. One major finding is how significant the females are in the groups (de Waal 1982). Males cannot

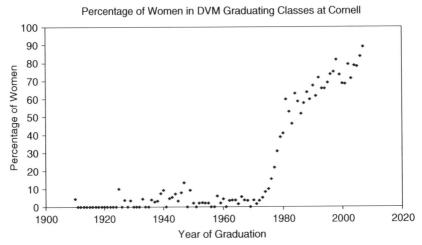

Figure 7.1 Percentage of women in DVM graduating classes, Cornell University. Reproduced by permission of Donald F. Smith, Cornell University.

achieve dominance without working with females, and females know this and play it to their own ends. In other words, intelligence is simply one of those things that both sexes need. Perhaps, as social animals unable to have their ways through brute force, females need it even more than males.

It is true that today the top-flight mathematicians tend to be male (and Asian or Jewish!), and that some professions heavy on mathematics and science have proportionately more males. There could be some biological factor – perhaps males needed great spatial abilities to go out foraging – but if there is, it is not great and the more likely cause is simply that females take other life routes. In particular, they balance work and families more (Ceci and Williams 2009). Also, interestingly, very talented mathematical females tend to be better at verbal skills than talented males. Does this imply that, because of their wider range of overall abilities, talented mathematical females have more career options than talented mathematical males? Could it be, almost paradoxically, that by directing females to use their mathematical talents we could be depriving society of the labors of those with high-level verbal skills? Either way, what is not clear is that in these respects Darwinian evolutionary theory is as sexist as the critics claim.

Even now you may feel that at most this amounts to – that is, women are being categorized by – a kind of catch-up solution, one which ultimately still assumes that women's place is *Kinder, Küche, Kirche* – children, kitchen, church. "What do women want?" asks a recent feminist critic of this whole line of thinking (Jordan-Young 2010). On the assumption that biology counts, the answer reduces to a simple response: "Women want men." Is there not some way out of this kind of thinking, once and for all? Don't just respond that (subject to all sorts of qualifications, many of which will occupy us in the middle part of this chapter) it is no less true that from the Darwinian perspective men are equally eager to have women. Can we do radically better, basically by making a break with Darwinism? No one now is suggesting that one drop evolution, but perhaps all of this talk about natural selection and adaptation can be tossed aside. This has certainly been an attractive option for a number of thinkers (including our critics), who often turn to the geneticist Richard Lewontin (2000) for support and illumination. He has already been introduced as a strong denier of ubiquitous adaptationism, a stance that he explicitly ties in with his enthusiasm for nineteenth-century strains of Germanic philosophy. Perhaps along these lines one can simply drop talk of adaptation and (while not denying a biological underbase) put everything interesting down to culture. So for instance male–female differences in academic achievement have little or nothing to do with biology and pretty much everything to do with opportunities and training, and so forth.

Well, of course, the answer is that one can take this route. The question is whether it is true. Marxism, to be frank, has not had an awfully good track record in the twentieth century, a point one can surely make while conceding that it nevertheless provides deep and important insights into the human condition. But even if one gets away from ultimate origins and motivations, and considers the question just on the surface – is talk of biological adaptation at all pertinent to or helpful in understanding human nature? – it is still a question that has to be answered and fleshed out. And although we shall be returning to these issues shortly, perhaps this is ultimately a question left to the reader. Earlier in this book, we saw that there is massive evidence that humans are part of the biological scene. We also saw evidence that the culture (or cultures) in which we are all immersed do demand new ways of thinking. Although this is not the main theme of the book, there is much to support the claim that, in the rest of the living world, Darwinian evolutionary mechanisms are the key to understanding change and the results of the process. If on metaphysical or religious grounds one persists in claiming that humans are different, that we are not covered by the same processes as the rest of organic nature, then there is little more to be said. Let us leave it at that.

Sexual orientation

We turn next to sexual orientation (Ruse 1988b). Two facts stand out. First, homosexual orientation is widespread and significant. The original Kinsey studies seem to suggest that the proportion of male homosexuals in the adult population of the USA is around 10 percent (Kinsey *et al.* 1948). In fact, a more nuanced reading of their results, confirmed by later studies many times in the USA and elsewhere, puts the figure lower but still striking, at around 3–4 percent – that is, about one in twenty-five men are people whose sexual desires are directed exclusively to members of their own sex (Laumann *et al.* 1994). The number of females generally comes in smaller, at about a third of this figure (Kinsey *et al.* 1953). More recent studies confirm these figures (LeVay 2010, 14). You might think that perhaps today (eighty years after the Kinsey studies) the numbers of lesbians would be higher, because women have more control over their sexuality; but although it is certainly the case that women seem more diffuse in their sexuality, the numbers are still far below the male numbers. Of course, being gay does not mean that one will not have children, but later Kinsey studies show that orientation (especially of males) is a very significant factor in having fewer offspring (Bell and Weinberg 1978). (Homosexuals are not infertile.) The second fact is that this is a problem for Darwinian theory. The emphasis is on reproduction. This does not, of course, mean that it is an insoluble problem. We have encountered reduced reproduction before. But it does mean that it is something that must be tackled. A disadvantage of even a few percent is over a small number of generations a devastating evolutionary handicap.

Analogously to the criticisms of the feminists, one might think that this shows that Darwinism is inherently prejudiced against gays, but this is surely not so. More obvious even than homosexuality is the fact that those most able to have sex together, namely siblings, in general (in both the human and the animal worlds) do not (van den Berghe 1979). The usual explanation, going back to the anthropologist Eduard Westermarck (1906), is that the incest barrier is in place so that very close relatives do not reproduce and that this is because severe inbreeding leads to horrendous genetic problems. The reason, obvious in the light of Mendelian genetics, is that many mutations have bad effects but that selection has worked to ensure that generally they do not show themselves. They are recessive to the selection-chosen normal genes (the "wild type"). However, inbreeding brings these bad recessive genes together (they are on both chromosomes) and sickness and death follow. Selection, therefore, has worked to make inbreeding less than desirable. It does this, not by giving you a gene to recognize siblings as such, but by giving you a biology that does not find those with whom you were in close contact while being raised sexually attractive. Expectedly, unrelated children raised together (as in the Israeli kibbutzim) also are loath to interbreed, and conversely one hears of siblings not raised together who feel attracted sexually. Whether or not one accepts this explanation, and most people do, it is hard to see it as inherently value laden, except perhaps against those who want to defy their biology. Similarly in the case of homosexuality, if Darwinism can explain it in some wise, one might in fact (if not necessarily) take comfort from Darwinism.

It is useful here to go back to that all-important distinction started by Aristotle: proximate causes, causes that bring things about, and final causes, causes that speak of goals or aims. (In his full version, Aristotle gives four kinds of cause, but he himself was prone to collapsing things down to two causes.) In dealing with sexual orientation, start with proximate causes and then move on to final causes. Darwinian theory is obsessed with design, with adaptation, so when we come to discuss final causes it is then that we expect to find evolutionary theory kicking in, although how and why it might kick in is going to be very much a function of what is found and concluded about proximate causes.

In dealing with proximate causes – what makes for sexual orientation – the big question is nature or nurture? Is one dealing with a cultural phenomenon or something biological or a mixture of the two? All will agree that culture has something to do with the case. It is very well documented that people confined to single-sex institutions – prisons, boarding schools, monasteries – often indulge in same-sex behavior. With reason one might object that this is not strictly to do with sexual orientation. Same-sex behavior can occur between people who are fully heterosexual in orientation and behavior when and if the constraints are lifted. But orientation per se can come into the mixture if the societal conditions are right. Contrary to popular belief, it is simply not the case that the ancient Greeks behaved like members of a San Francisco bath house before the AIDS epidemic (Dover 1978). Anal intercourse, for instance, was looked down on and could be subject to humiliating penalties (including ritualistic shaving of the pubic hair, a practice of Greek women, and thus considered demeaning for men). However, upper-class young men had homoerotic passions and liaisons, and these – especially between an older youth and a younger boy – could eventuate in a sophisticated form of masturbation. ("Intercrural," where the older would take the younger on his lap and stimulate himself on the younger's thighs.) It is known that there was some lesbian activity, although much less is known because it was something of a taboo subject (like menstruation). These sorts of passions and activities were known also in later times, famously (or notoriously) in British public (meaning private) schools.

It was generally the case that adult Greeks were heterosexual. However, as the Socratic dialogues show, homosexual passions could persist. It is thought that Plato might have been homosexual in inclination and behavior all his life – in our language he was "gay," although it is a little anachronistic to use it of people back then. The existence of homosexually oriented people down through the ages has been well documented by historians (Boswell 1980). There are for instance a couple of men we would call homosexual in that thoroughly heterosexual romp, the eighteenth-century novel *Fanny Hill*. This incidentally gives the lie to people like the late French historian-philosopher, Michel Foucault (1978), who argued that homosexuality did not exist before the nineteenth century, when it was invented as a power play by the medical profession to control certain people. It is not a pure "social construction," although it is true that it was increasingly medicalized in the nineteenth century and that the term "homosexual" was not invented until the middle of that century. "Homo," incidentally, refers to the Greek for "same" rather than the Latin for "man." Technically the first vowel should be short, as in "hot," not long as in "home."

Freud

In the first part of the twentieth century, there were those (often psychoanalysts assuming the mantle of Freud) who thought that all aspects

Figure 7.2 Ancient Greek vase showing lesbian activity, Tarquinia National
Museum (Italy).

of sexual orientation are environmental. But this belief has faded – perhaps the most dramatic counter-evidence is of children with one kind of
set of sex chromosomes brought up as the other sex (as the famous case
of a boy whose penis was destroyed by circumcision), but who in their
teens asserted their commitment to the gender of their chromosomes, and
with this a totally heterosexual orientation. Freud (1905) himself always
believed that in some cases sexual orientation is simply given by nature,
by biology, although he could suggest no reason for this. However, he was
certainly a major force in the drive to find or ascribe significant cultural or
environmental factors in the formation of sexual orientation. One hypothesis Freud floated, using Greek homosexual activity as prime evidence, was

that sexual orientation is a function of imitation. You think and behave as an adult according to the ways in which you were shown or tutored as a child or youth. Another hypothesis, the one that really caught people's imaginations, rested on Freud's theory of the Oedipus complex and how it can backfire, through the "dysfunctional triangle."

Setting the scene for virtually all of the thinking in the twentieth century about etiologies (proximate causes) – nature and nurture – Freud saw a continuum between male and female. These categories are not one or the other, like triangles and circles. In human development (for males that is, Freud always tacked females on at the end, if at all), we start heterosexually fixated on mother and the breast; we then turn homosexual, fixated on our own feces (the first things we have produced on our own) and our genitals; at some point around four or five this all goes underground; then at adolescence we wake up again and move over to a permanent state of heterosexuality. But to make this last move, we must give up our love of mother (Oedipus' state) and transfer our affections to other females. Some, however, cannot do this, and revert back to a homosexual phase, permanently. (Homosexuality for Freud therefore is an immaturity, not a sickness.) The reasons for reversal are diverse, but the most common is a family with an overbearing or hostile father and a too-nurturing mother. On coming to sexual awakening, one wants to have sex with mother, but one realizes that one cannot have sex with her. This violates the incest taboo (something Freud thought, in Lamarckian fashion, is engrained in the human species from past events when brothers castrated their father and then made a compact not to have sex with mother). Furthermore, such sexual activity will arouse the hostility of father. So one reverts to the safer option of having sex with no woman, only men.

Studies have shown repeatedly that there is something to all of this, at least at the surface level (Bell *et al.* 1981). Male homosexuals are more likely to report dysfunctional families of the kind just described. However, as many have pointed out, perhaps Freud got it all backward. Could it be that a person who is homosexual in adult life has elicited certain behaviors in the parents when he is young? There are very good studies showing that future homosexuals are more likely to engage in cross-gender play when children – what in the old days (before we became more sensitized to language use) were called "sissy boys" and, relatedly, tomboys for girls (Green 1974). There are more future-heterosexual tomboys than sissy boys, perhaps because being a tomboy does not have the social stigma. Could it be that future gays (men) alienate fathers with their behavior and encourage mothers to be overly protective? In this case, then, since the environment comes second, could biology come first? By the middle of the last century, some researchers were starting to think so, especially given the increasing unease with Freudianism as a system of therapy and its failure to give a serious causal basis to its ideas – the Lamarckian underpinnings of the Oedipus complex being but one instance (Sulloway 1979).

Biology (at the proximate level)

Increasing understanding of human development encouraged the biological approach to sexual orientation. It was apparent that to make a human you start with the female (Money and Ehrhardt 1972). All other things being equal, a zygote will develop into an adult female. To get the male, you add things, namely crucial hormones. This brings about, most obviously, the growth of the male sex organs. (Chromosomal males who are insensitive to the hormones develop as females – usually not quite all of the way, because they are only insensitive to some hormones and not to others.) Could it be that sexual orientation is simply a matter of the balance or imbalance of hormones, a matter of the levels of the hormones that make for males and those that make for (or keep as) females? One or two things became clear pretty quickly. First, that if it is a matter of hormones, then it is going to be a matter of delicate balances. As noted earlier, homosexuals generally have no issues at all with gender identity. Whatever is happening, not only are the sex organs being produced, but so also is a parallel sense of having the right gender appropriate to the sex organs. This of course does not preclude gay men having other features that tip them more (than heterosexual men) toward the female end of the spectrum, and conversely for lesbians. As it happens, there are huge numbers of studies confirming this suspicion – not for every gay man or lesbian on every occasion, but systematically and strongly. These include cognitive traits and personalities. There really are more gay male ballet dancers than one might expect. Second, talking of the putative effects of hormones, it was not simply a matter of hormones in the adult. Early workers (often working on less than entirely voluntary subjects) soon found that upping the levels of male sex hormones in males and comparably in females simply increased the sex drive in the already-acquired orientation. It did nothing to change directions or inclinations.

The move then was to go back, even to the developing fetus. Perhaps this was the place for the real action. In the 1960s and 1970s there was some very suggestive work on other organisms, rats in particular (Dörner 1976; Ruse 1988b). Copulating rates show very distinctive patterns of behavior, specifically with the female going into a crouching pattern, called "lordosis." By castrating rats at suitably early ages and by manipulating hormonal levels it was possible to reverse the male–female behavior entirely, with females mounting and males crouching in the lordosis position. It should be said that by the 1960s, people were feeling increasingly uncomfortable with labeling homosexuals sick or deviant – test after test on people in normal circumstances showed no significant mental health issues differentiating straights from gays. This perhaps led to a less than enthusiastic and immediate reception of the early hormonal cause thesis, because the main worker on the subject, Gunter Dörner, was not only a scientist working in the German Democratic Republic (East Germany), which was bad enough in itself, but he explicitly labeled homosexuality a sickness and had the avowed aim of reducing or eliminating it. (He changed his mind later, seeing homosexuality simply as a value-free variant.) However, the evidence did keep coming in, and more and more people started to think that hormonal levels in early development might be very significant – specifically researchers focused in on the hypothalamus, which develops in humans between the third and sixth month of fetal development. Could it be that relative hormonal levels at this time of growth were significant? It seemed so, and still seems so. Following this up, there has been a huge amount of work to find out if one can actually find physical differences in the brains of gays and straights, and if so whether these suggest moves along the male–female scale, something one expects on general grounds of personality and abilities, quite apart from objects of sexual interest. There have been some positive findings, including one much-publicized study claiming that the neurons of the hypothalamus differ from male to female, and from straight to gay, and that gay men fall more toward the female end of the spectrum than do straight men. As so often with these studies, however, there are unanswered questions and issues – for instance, the study in question was of gay men most of whom had died of AIDS-related illnesses. Was this a significant factor distorting the findings?

Probably not, a belief supported (although not definitively proven) by later studies. (See Wilson and Rahman 2005; Swaab 2008; and LeVay 2010 for more details.)

If we are going to go the biological route, what about starting with the genes? The most prominent route for teasing out proximate causes is twin studies, where one compares monozygotic twins (from the same egg and thus genetically identical, that is 100 percent) with dizygotic twins (different eggs and no more related than normal siblings, that is 50 percent). Two results stand out (Herschberger 2001). First, monozygotic twins are much more likely to share sexual orientations. It could be all put down to environment, since monozygotic twins are supposedly more likely to be treated the same than dizygotic twins, but there are ways of teasing this apart and the answer is that it is unlikely to be a purely environmental effect. Nevertheless, one does find indubitable cases of monozygotic twins with different sexual orientations and so it is simply impossible that it is genes and only genes at work. Some kind of hormonal factor (if not the only cause) is not ruled out, because some monozygotic twins are in the same amniotic sac and some are not. There could be different pertinent pre-birth influences even for monozygotic twins.

Expectedly, heroic efforts have been expended to find "gay genes" (Hamer and Copeland 1994). Some intriguing findings suggest that gay men are more likely to have gay male relatives on the mother's side of the family than the father's side (Hamer *et al.* 1993). Since men get their X chromosomes from their mothers (and Y from their fathers) this suggests that the place to search might be on the X chromosome, and there are some positive findings. More research suggests, however, that it is unlikely that any one gene causes orientation. A number will be involved. Note also that because genes are involved, it does not follow that environmental factors are irrelevant, or that even with a certain gene combination all children will be affected the same way. One fairly solid finding (reached through a meta-analysis of other studies) is that birth order can be significant (Blanchard 1997). Brothers lower down the birth order (more precisely, men with more older brothers) are more likely to be gay, and this seems to be connected to biology because the finding does not hold when one is dealing with adopted children. It is thought that the cause might be that male offspring cause an antigen immune reaction in mothers, something that gets stronger with the birth of each male child. Biology is working here, obviously. There could be a significant genetic factor, inasmuch as only certain mothers with certain genes are going to build up these immunities.

Biology (at the ultimate level)

What happens when we move away from proximate causes to ultimate or final causes? As noted, even a 3–4 percent ratio does not happen by chance, at least not in the Darwinian world. Drift is very unlikely. In addition, as noted earlier in the chapter, Joan Roughgarden (2009) points out that same-sex activity of some sort or another is common throughout the animal world, although most would doubt that it is the group-bonding phenomenon that she supposes. Drawing on what we have seen already, one obvious suggestion is that homosexual orientation is a function of balanced heterozygote fitness – gays are those that do not reproduce, balancing out siblings who reproduce more than average. Another suggestion is that gays are "helpers at the nest" – for whatever reason, they are not going to reproduce themselves (or not reproduce as well themselves) and so they don't reproduce at all but help close relatives in their reproduction. A variant on this could involve something known as "parental manipulation," where parents push one or more offspring into non-reproduction, in order to raise the reproductive chances of other children (Alexander 1974).

Is there any evidence for any of this? Certainly nothing that is overwhelmingly definitive. Some hypotheses have not borne fruit, at least not yet. For instance, researchers have failed to find direct evidence that homosexual men are more likely to aid close siblings than heterosexual men. However, there are some positive straws in the wind. Most particularly, there is a large study that links homosexual males with increased fecundity on the mother's side, and interestingly fails to find such a connection on the father's side (Camperio *et al.* 2004). This suggests that the decreased reproduction of the gay son is balanced by increased reproduction in the rest of that side of the family. The biology suggests that what is happening is increased attraction in some sense to males – in men this leads to homosexual orientation and in women to more offspring.

The older-brother effect is more puzzling. The obvious Darwinian explanation is that having different kinds of sons is a form of parental manipulation, an adaptive strategy of mothers. (Note that there is no implication that such a strategy would be known to mothers.) Perhaps having older sons of one kind and younger sons of another kind makes good evolutionary sense. There is some evidence that having slightly feminizing features can be an attractive feature in (heterosexual) males – metrosexuals really do have more fun! It has been suggested that as one goes down the birth order these features build up, perhaps eventually tipping over into homosexual orientation (Miller 2000). Remember Hamilton's (1967) theorem about "local mate competition." It is based on the insight that if brothers are competing for the same females, from a parental perspective it makes no difference which one succeeds and hence it makes more evolutionary success to reduce the number of male offspring, biasing the sex ratio in favor of females. The case of homosexuality could be something similar, where, given the abilities of humans to adopt different behaviors, rather than avoiding males it is biologically worthwhile to give these behaviors a shove in (maternally desired) directions, even at the expense of producing some (more or less) non-reproductives.

As with male–female differences, there are those who feel that there is something suspect about the whole Darwinian approach to sexual orientation. (These two groups of doubters often have overlapping memberships.) However it is caused, such critics prefer to think in terms of cultural or environmental factors. But as before, one has to ask whether such a stance is taken solely on ideological grounds or if there is solid empirical or theoretical evidence backing such a stance. For instance, the critic mentioned in the context of the earlier discussion of male–female differences hardly inspires great confidence, for in her discussion of sexual orientation she avoids any mention of many of the more interesting (and surely pertinent) facts about the topic, such as childhood behavior and attitudes, birth order, and so forth (Jordan-Young 2010). What she does do is describe all studies on brain function as "quasi experiments" as opposed to "true experiments," because obviously we cannot go in and manipulate people simply to gain information. In the philosophical trade we refer to this as offering a "persuasive definition," meaning that one is trying to tilt things a little while ostensibly simply clarifying issues. As before, the reader will have to judge whether the work discussed above is essentially second-rate, as this critic implies. And as before one has to decide if one is comfortable with an ideology that drives one to make an exception of human beings with respect to the forces that mold all other organisms, and that were certainly important in our own evolutionary histories. No doubt most of the critics of Darwinism are thoroughly secular, but is there perhaps some overlap with the forces that drive many religious people to make an exception for humankind?

Ethical implications?

Finally, even if any of these causal speculations are partially or mainly true, what implications if any does this kind of work have? From a philosophical viewpoint, apart from epistemological issues like testing and so forth, it may have relevance for certain moral theories. Utilitarians from Jeremy Bentham on have had relatively little trouble with homosexuality from a moral perspective (Ruse 1988b). So long as it does not harm others, and there is no evidence that homosexuality is linked to pedophilia, for instance, then it is acceptable. Kant did not much like it, arguing in a convoluted way that one was not treating the sex objects as an end in themselves but just using them for one's own sexual gratification. Inasmuch as this is a valid point, then presumably this applies to heterosexual behavior also, but conversely if heterosexual behavior can be allowed morally (on the grounds that one gives oneself to the other as one takes from the other) then there seems no good reason why this should not apply also to homosexual behavior. Either way, biology does not seem to contribute much to the discussion. When one gets to Christian ethics, especially Roman Catholic ethics, which makes much of Aquinas's doctrine of natural law, the science could start to kick in (Ruse 2010). For the natural law theorist, what one should do is that which is natural, because that is how God has made things and intends them to be used. Artificial contraception for instance was barred because it went against the way in which God had designed reproduction and was supposedly bypassing the real reason for sexual intercourse. One was just having sex selfishly for pleasure. Homosexuality, especially male anal intercourse, likewise got St. Thomas's condemnation. Penises are made for reproduction and anuses are made for defecating. That is all there is to the matter.

> We have said that God exercises care over every person on the basis of what is good for him. Now, it is good for each person to attain his end, whereas it is bad for him to swerve away from his proper end. Now, this should be considered applicable to the parts, just as to the whole being;

for instance, each and every one of his acts, should attain the proper end. Now, though the male semen is superfluous in regard to the preservation of the individual, it is nevertheless necessary in regard to the propagation of the species. Other superfluous things, such as excrement, urine, sweat, and such things, are not at all necessary; hence, their emission contributes to man's good. Now, this is not what is sought in the case of semen, but, rather, to emit it for the purpose of generation, to which purpose the sexual act is directed. (Aquinas 1975, III, 122, 143)

To use the technical language, homosexual acts are "unnatural vices," *vitae contra naturum*. This was reaffirmed by Pope Benedict XVI, speaking when he was still Cardinal Ratzinger on behalf of the Congregation for the Doctrine of Faith to the Catholic bishops: "To choose someone of the same sex for one's sexual activity is to annul the rich symbolism and meaning, not to mention the goals of the creator's sexual design" (reported in *The Times*, October 31, 1986). That is to say, God is not arbitrary. He is behind the moral standard but He offers a package deal – morality, behavior, physical nature.

If you want to criticize this on its own terms – and you might not want to do this – then you cannot simply ignore what is natural. You cannot simply say: if it feels OK, then it is OK. You have got to show that the behavior (or feelings) you are defending is natural. You might do this, of course, by arguing that what is "natural" for humans is not necessarily what is natural biologically. It is natural to read a book or to farm a field, but goes beyond biology in both cases. However, when it comes to our own bodies, which do have natural (biological) functions, I suspect the natural law theorist is going to insist that these be taken into account. You cannot just ignore them in the name of culture. But in both cases, contraception and anal intercourse (and homosexual behavior and inclination generally), modern evolutionary biology seems to be on the side of the revisionist. We now know that sexual intercourse in humans, as in the bonobos, does more than simply produce babies – and was cherished by selection for more than producing babies. In the bonobo case, it is for general bonding. In the human case, it is at least for promoting bonds between mates – keeping the relationship going, and most importantly keeping males involved and willing to contribute to parental care of the offspring. If you take also the flip side, that apparently it is natural – or at least acceptable to the moral law theorist – to tackle childhood diseases so that far more children grow to maturity than would otherwise, and that producing a maximum number of children (all requiring parental care) is not necessarily the best thing biologically, the biological case for artificial methods of contraception follows readily.

Likewise for homosexuality. If indeed it is something that follows as a consequence of natural selection rather than against selection, then the case for saying that it is part of the natural biological state of affairs follows at once. Do you want to say nevertheless that it is a misfortune, and that the gay person is like the sufferer from sickle cell anemia, only existing because others are benefiting? (See the next chapter's discussion of sickle cell anemia for more on this topic.) That is surely something that needs more assumptions, not all of which are obviously true. It could be that being gay or lesbian is adaptive, if some kind of kin-selection model did actually apply. It could be that it is non-adaptive for the individual if some kind of balanced fitness model applies. It could be that it is non-adaptive for the individual, but adaptive for the parent (or at least a reproductive cost the parent can bear). But being non-adaptive does not necessarily make something non-natural, even for a Darwinian. In certain respects, the whole point is that it can be very natural. Otherwise, why bother with the explanatory models? On a different level, note that being gay is not something that necessarily makes you miserable. It follows then that, apart from the genetic diseases that do make you miserable, you cannot really say that gay or lesbian orientation is something non-natural because God would always want (and be able) to make us happy. Basically, if you believe that God stands behind the whole creation, the case for saying that some sexual orientations are immoral or unfortunate seems more and more tenuous. (There will be more in the next, final chapter on general issues to do with health and sickness.)

What is race?

We turn finally to the third topic, namely race. Here the reference is to a distinguishable group of human beings, differentiated in their biology. You do not have to go far to see why this is a topic that makes a lot of people very tense.

> Not only physical traits, like eye color, skin color, body build and such characters as stature, color and form of hair, proportions of racial features

and many others are inherited in race-crosses but also mental traits. This is a matter which is often denied, but the application of methods of mental measuring seems to have produced indubitable proof that the general intelligence and specific mental capacities have a basis and vary in the different races of mankind. Thus it has been shown, by standard mental tests, that the Negro adolescent gained lower scores than white adolescents and this when the test is made quite independent of special training or language differences and also when the children tested have a similar amount of schooling. (Davenport 1930, 557; quoted by Kitcher 1999, 231)

This was the USA in the first part of the twentieth century. Throw in Hitler and National Socialism, and you can see why the very mention of race made some people very tense and eager to deny that it has any meaning or application at all. The well-known anthropologist Ashley Montagu wrote a book with the revealing title *Man's Most Dangerous Myth: The Fallacy of Race* (1942), and he was one of those involved in a declaration on the topic in 1950 by UNESCO.

For all practical social purposes "race" is not so much a biological phenomenon as a social myth. The myth of "race" has created an enormous amount of human and social damage. In recent years it has taken a heavy toll in human lives and caused untold suffering. It still prevents the normal development of millions of human beings and deprives civilisation of the effective co-operation of productive minds. The biological differences between ethnic groups should be disregarded from the standpoint of social acceptance and social action. The unity of mankind from both the biological and social viewpoints is the main thing. (Montagu *et al.* 1950, 14)

And yet, today, a lot of people are not quite so sure. Indeed we make social policy on the basis of race, intentionally, through such methods or programs as affirmative action trying to do good things. Analogously we find that the medical profession often asks questions about race, believing that this can be significant in the treatment of and (even more) the finding of biologically caused ailments. For instance, the urological group that I use here in Tallahassee sometimes arranges free screening for prostate cancer. It puts out a special call to the black (African American) population, believing it to be at special risk. The questions therefore are whether race has any real biological meaning; why and how and whether evolution is involved somewhere; and what the implications (including social and moral) might be.

The idea of "race" goes back at least to the ancient Egyptians, if not earlier. Although the different suggested definitions or characterizations were all over the place, suggesting a significantly subjective factor to the notion, no one seems to have had any real qualms about the reality of the idea. Nor was it seen as at all morally objectionable, and indeed in medieval times was often linked to the biblical story of Noah and his three sons, Shem, Ham, and Japheth. Black people were thought to be the descendants of Ham, who was cursed for mocking his father when he lay naked and drunk. By the time of the Enlightenment, the discussion was sharpening a little as well as being fleshed out with considerably more knowledge of the different peoples around the globe. Linnaeus set the background with his system of classification, insisting that organisms be put in a series of nested sets (known as taxa), each level (known as category) being higher and more comprehensive than the one below. (See chapter 1.) Remember that humans are included in this system, starting at the top (the kingdom category) in the overall taxon known as animalia, and working their way down to *Homo sapiens* (species category). (See chapter 2.) Often species were divided into groups, subspecies, and the notion of race was often incorporated within this level of category. General opinion was that taxa at the level of species are in some sense real or objective, and that this is no less certain for taxa at higher levels or at the subspecies level either. One of the most influential (on Kant among others) attempts at offering a racial classification was that of the German biologist Johann Friedrich Blumenbach, who in his 1775 tract *The Natural Varieties of Mankind* offered a fivefold division: the Caucasoid, Mongoloid, Ethiopian (better known as Negroid), American Indian, and Malayan. Jumping ahead a century to the time when Darwin was just doing his work on humans, Thomas Henry Huxley (1870) offered a slightly more detailed grouping, putting us into nine groups or races (see Figure 7.3).

In the *Descent of Man*, Darwin was very interested in the question of human races and devoted considerable time to the topic, although he did stay out of the naming and classifying debate. He was convinced, on the basis of shared physical and mental features, that we humans are all descended from one group. We are one species. (He was a "monogenist"

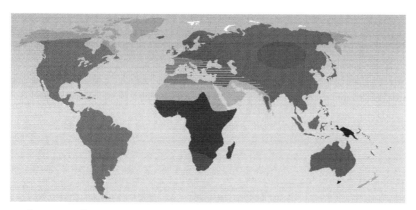

Huxley's map of racial categories from *On the Geographical Distribution of the Chief Modifications of Mankind* (1870).

1: Bushmen		7: Polynesians
2: Negroes		8: Mongoloids A
3: Negritoes		8: Mongoloids B
4: Melanochroi		8: Mongoloids C
5: Australoids		9: Esquimaux
6: Xanthochroi		

Figure 7.3 Thomas Henry Huxley's racial classification.

rather than a "polygenist.") He thought that there are differences between races:

> There is, however, no doubt that the various races, when carefully compared and measured, differ much from each other, – as in the texture of the hair, the relative proportions of all parts of the body, the capacity of the lungs, the form and capacity of the skull, and even in the convolutions of the brain. But it would be an endless task to specify the numerous points of structural difference. The races differ also in constitution, in acclimatisation, and in liability to certain diseases. Their mental characteristics are likewise very distinct; chiefly as it would appear in their emotional, but partly in their intellectual, faculties.
> (C. Darwin 1871, I, 216)

Darwin revealed the ambiguities of a Victorian liberal. If the real mark that one has a species, and that that species represents something clearly objective, is that there is reproductive isolation between its members and all other organisms, then he was not at all convinced that this criterion holds between human groups. Having said this, however, on one matter Darwin was unambiguous. For all that other races are human and deserve dignity for that reason (including being free from slavery), the European race is by far the superior, especially mentally.

> The belief that there exists in man some close relation between the size of the brain and the development of the intellectual faculties is supported by the comparison of the skulls of savage and civilised races, of ancient and modern people, and by the analogy of the whole vertebrate series. Dr. J. Barnard Davis has proved by many careful measurements, that the mean internal capacity of the skull in Europeans is 92.3 cubic inches; in Americans 87.5; in Asiatics 87.1; and in Australians only 81.9 inches. Professor Broca found that skulls from graves in Paris of the nineteenth century, were larger than those from vaults of the twelfth century, in the proportion of 1484 to 1426; and Prichard is persuaded that the present inhabitants of Britain have "much more capacious brain-cases" than the ancient inhabitants. (I, 145–46)

Darwin also had little doubt about what would happen in the future. "At some future period, not very distant as measured by centuries, the civilised races of man will almost certainly exterminate and replace throughout the world the savage races" (I, 201). Although he did suggest that Western diseases will do the job at least as efficiently as anything else, if not more so. You can see here, incidentally, why the question of Hitler and evolution is complex. On the one hand, there is no question but that the Nazis inherited a nineteenth-century legacy of racism. On the other hand, while you may complain that the evolutionists went some way to legitimate it, it is clear that the evolutionists took it in from general cultural thinking, rather than invented it for the first time. Moreover, someone like Darwin is far from preaching some kind of world struggle with whites going out and vanquishing and destroying all other races. And there is nothing here to support Hitler's obsession with the Jews.

Why are there different races? Darwin thought that natural selection plays some role, and also (as always) Lamarckian use and disuse had a role to play. But as we know, it was sexual selection that Darwin saw as really important in the formation of human characteristics, and he was quite convinced that most if not all of the distinctive features separating races is a result of this process. It is all a matter of different standards of beauty and so forth. Remember the Hottentots. Remember also that a popular view at the time (Spencer held it) was that the white races are brighter than the darker races because they have lived in tougher conditions and hence had to use their wits more. They are not as con-genitally lazy, either. This is really not the way that Darwin was going. Whites are brighter because that is what they wanted in mates. It is as simple as that.

The whites-had-to-work-harder thesis had long legs, running well into the twentieth century, and was given a selection-based explanation. The key work pushing this idea was by W. D. Matthew, a paleontologist at the New York Museum of Natural History.

> We should expect ... to find in the land life adapted to the arid climatic phase a greater activity and higher development of life, special adaptations to resist violent changes in temperature and specializations fitting them to the open grassy plains and desert life. In the moist tropical phase of land life, we should expect to find adaptations to abundant food, to relatively sluggish life and to the great expanse of swamp and forest vegetation that should characterize such a phase of climate. (Matthew 1915, 173)

Even as late as 1939 this essay was republished as a book. Among others who responded favorably was the then-former president, Theodore Roosevelt. He was "intensely interested" (Colbert 1992, 108–109).

Move the clock forward now from Darwin toward the present. Apart from the political issues, a major reason why the notion of race was thought less significant than it was for someone like Darwin was because of newly revealed genetic evidence. It appears that the variation between human groups is significantly less than the variation within such groups. The Harvard geneticist Richard Lewontin (1982) proclaimed that 85 percent of such variation was within and only 15 percent was without. It is true that there are some highly visible features, and Lewontin himself was not about to deny these – skin color and morphology, for instance – but liter-ally as well as metaphorically these are skin deep. (See Figure 7.4.) Another recent major study of human variation, looking at 1,056 individuals from 52 populations, backs this: "Of 4199 alleles present more than once in the sample, 46.7% appeared in all major regions represented: Africa, Europe, the Middle East, Central/South Asia, East Asia, Oceania, and America. Only 7.4% of these 4199 alleles were exclusive to one region; region-specific alle-les were usually rare, with a median relative frequency of 1.0% in their region of occurrence." Put things another way: "Within-population dif-ferences among individuals account for 93 to 95% of genetic variation;

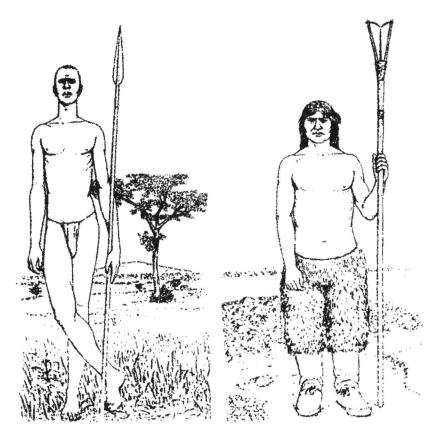

Figure 7.4 African versus Eskimo. The different shapes are thought to be adaptations, to hot and cold climates, respectively. From Lewontin (1982), in turn reproduced from Howells (1960).

differences among major groups constitute on 3 to 5%" (Rosenberg *et al.* 2002, 2381). All of which is to be expected if humans only came out of Africa 150,000 years ago or later, going through a population bottleneck at some point during that time. Hence: "Human racial classification is of no social value and is positively destructive of social and human relations. Since such racial classification is now seen to be of virtually no genetic or taxonomic significance either, no justification can be offered for its continuance" (Lewontin 1982).

However, Ronald A. Fisher's last student, Anthony Edwards (2003), points out that a fallacy lurks here. It is true that if you look just at one locus, and only at the alleles at that locus, then between two groups there might be little genetic distance. But once you gather in information from different loci and put it all together, if the variations are correlated, you get a very different story. Pertinently, then groups do start to emerge strongly. This was just what was found in the recent study mentioned above. Correlation does kick in. If you run a cluster analysis across the large sample, the differences are sufficiently strong that the genetic information puts people into groups that correspond to ethnic sortings. For instance, geographic Europeans come out as one genetic cluster and Africans come out as another genetic cluster. One can go further. Factoring in more and more genetic information yields ever-finer divisions, and in parallel these map ever-finer ethnic and geographical groups. Nor is this a simple artifact. Work with different amounts and samples of genetic information and you get the same findings. By way of example, the analysis picks out as anomalous a group in northern Pakistan. These are the somewhat isolated Kalash. Tradition has it that they are not of the same ethnic background as the rest of their countrymen, but rather of European or Middle Eastern origin, a finding that genetics backs. In short: "Genetic clusters often corresponded closely to predefined regional or population groups or to collections of geographically and linguistically similar populations" (Rosenberg *et al.* 2002, 2384).

Are races real?

Philosophers have suggested two different ways in which one might move forward from here, trying to give some kind of real objective reality to races. One is historical or phylogenetic. The other works with the evidence of evolution from shared ancestors. Relying on material using various standard techniques for inferring phylogenies ("cladistics"), Robin Andreasen offers the following phylogeny of human groups going back to the shared ancestors (see Figure 7.5). She argues that it is through such a tree as this that we can see the way forward to "give a biologically objective definition of race" (Andreasen 1998, 214; see also Andreasen 2000). Fellow philosopher Philip Kitcher, who seems to endorse a similar position – "*The concept of race is a historical concept*" (Kitcher 1999, 236; see also Kitcher 2007 and Hardimon 2003) – argues that although clearly today we are seeing a genetic breakdown across the racial boundaries, even here things might be more stable than one expects. On the one hand, a little bit of interbreeding at the edges of a population might not make that much difference to the population as a whole. "Studies of the history of marriage in southern

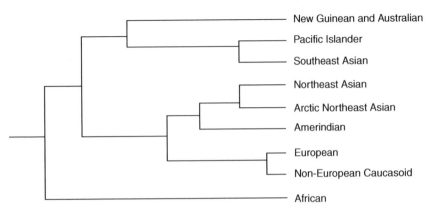

Figure 7.5 Human phylogeny.

England and in Italy testify to an amazing proximity of spouses, even comparatively recently" (Kitcher 1999, 239). You are not going to get a lot of gene flow across a group. On the other hand, even where there is significant mixing, as in the USA, racial borders are not necessarily crossed that often when it comes to breeding. Noting that black–white marriages in the USA in 1970 tended to be less than 1 percent of marriages, Kitcher notes that even though it may be physiologically possible to interbreed, there is nevertheless a significant degree of reproductive isolation. Behavior and preferences can be just as important as straight biology in these cases. "Specifically, if the 'blacks' and 'whites' in a particular region at a particular time reproduce together at a relatively low rate, then we can say that there is an incipient racial division between those groups at that place and at that time; if the rate of inter-reproduction remains low across a period, then we can talk about two races in that region" (Kitcher 1999, 240).

Note that there is no talk here about adaptation, although presumably if Kitcher's thinking is well taken, we do have sexual selection at work. Blacks prefer black partners and whites prefer white partners, and it is probably not just social factors at work. (Although social factors give you sexual selection, too.) In fact, Andreasen's analysis rather points away from adaptation, because one of the main techniques for inferring phylogenies relies on molecular genetic drift. Over time, the molecules of two groups drift apart at a rate that can be quantified. If natural selection is at work, then this spoils the picture. Notice also that there is a certain amount of revision of terms at work here, especially in Andreasen's proposal.

Normally we think of races as being fairly limited in number: white, black, Asian, and so forth. The numbers vary, but we think in terms of ten or less. Huxley (1870), for instance, had nine, although "Mongoloid" was divided into three. A cladistic analysis is liable to be much more fine-grained. So also is the Rosenberg analysis, referred to above. Philosopher Joshua Glasgow (2003) thinks this is fatal to Andreasen's case. He writes that "this exposes what I take to be the central flaw of the cladistic approach. That is, Andreasen has found a way of carving our ancestors into breeding populations, but these populations are not what we call 'races'" (459). In response, however, Andreasen might agree but then go on to point out that this is generally the whole point of a scientific approach to what hitherto has been accepted as common sense. One does come up with finer divisions based on more complete understanding. There is nothing stopping someone taking the fine divisions and then grouping these into more inclusive sets. If this proves impossible or does not correspond at all to folk notions of race, then perhaps indeed the very concept crumbles and should go. In fact, however, this seems not to be the case. In their own words, Noah Rosenberg and his co-workers "identified six genetic clusters, five of which correspond to major geographic regions" (Rosenberg *et al.* 2002, 2381), adding "and subclusters that often correspond to individual populations."

Another approach to race, promoted by philosopher Jonathan Kaplan and biologist-turned-philosopher Massimo Pigliucci, does appeal directly to adaptation. Denying Andreasen's claim that you can differentiate groups on the basis of phylogeny, they make adaptation all-important in their concept of race. "As long as differences between populations can be maintained because of their adaptive significance, races can persist despite significant gene flow between populations" (Pigliucci and Kaplan 2003, 1165). Unfortunately the two authors are not terribly helpful in telling us exactly what they have in mind for races under their terminology, or in giving examples. One presumes, for instance, the two individuals pictured by Lewontin would count as members of distinctive races, since many of their features are surely adaptive – skin color, for instance (light skin is now believed to be connected to the adaptive advantage of getting sufficient vitamin D when the sun is relatively weak), and body shape, for another (the northerner has a body well designed for the cold, minimizing skin surface against weight, for example). The connection between the more traditional notions of race is apparently there but weak.

> While it seems clear that biologically meaningful races will not
> correspond well to folk racial categories, this does not imply that folk
> racial categories are completely orthogonal to biologically meaningful
> racial categories. However, insofar as there is evidence that biologically
> significant human races do exist, that evidence points toward *most*
> biologically meaningful human races being quite a bit smaller and far
> more numerous than are folk races; the idea that those groups picked out
> by folk races and those populations forming biological races will not, in
> general, correspond is therefore likely correct. (Pigliucci and Kaplan 2003,
> 1166)

The two authors are fuzzy on what behavioral or intellectual factors might distinguish groups. Edward O. Wilson (1978), writing in a similar vein, has suggested that the fact that Asians apparently are less able to digest alcohol than people from the West might lead to different social behaviors. Perhaps something similar might be said of the ability to digest cow's milk, something we have noted earlier as a recently acquired adaptation and not one possessed by all peoples of the world. Parenthetically, the non-correspondence between their understanding of race and folk understandings seems to be taken by Pigliucci and Kaplan as a point in favor of their analysis. No one can accuse them of simply trying to justify morally offensive, all-too-prevalent cultural notions.

It was William Whewell who saw that objectivity in classification is essentially the same thing as truth in theories, namely something marked by a consilience. (See chapter 2.) We tend to think a classification objective or real if it is not just arbitrary but, in Plato's language, carves nature at the joints. A classification shows its objectivity by having more than one way of making the same division. Making a library classification just alphabetically is useful but not objective. Classifying a gem is objective, because the physical features correspond to the molecules – both routes lead to the same divisions (Ruse 1991). Many philosophers think this applies in biological classification. Species (taxa) are thought real because a division made on the basis of reproductive isolation corresponds to a division made on physical or behavioral features. Humans look alike and they don't have sex with cows. Many have rejected the subspecies concept precisely because you don't get the consilience – divisions one way do not correspond to divisions made other ways. This really is the epistemological issue with races. Someone like Andreasen is trying to show that history corresponds to the genetic distributions that we have now – the two classifications coincide, and I take it that Kitcher is trying to add behavior, too (or at least reproductive isolation in a sense). Apparently, Rosenberg and his co-workers would also throw in geography and language. Pigliucci and Kaplan deny that one can get such a consilience in this direction, and so they are trying to show that you can get one via genetics and via adaptive functioning. These particular groups are separate, and they function (or rather their members function) in the same distinctive ways. There is some overlap in the two sets of results (one presumes so, given the coyness of Pigliucci and Kaplan) but no one pretends to offer a confident overall synthesis. Either way, however, evolution has been important, although perhaps for Pigliucci and Kaplan Darwinian evolution is more important than for Andreasen and Kitcher. Either way, notice that we have gone far beyond the counter-reaction of someone like Ashley Montagu. There seems to be agreement that there is something to the notion of race and this is not purely social or cultural. In some sense, races are real or objective.

Is racial thinking immoral?

Finally, what if anything can be said at the social or moral level? Race matters, all agree on that. Race (however understood and however the divisions are made) has a biological component. This does not deny a social dimension. Kitcher, in particular, is good on this, pointing out that race is something surrounded by cultural beliefs and prejudices and practices. No one living in America could possibly say that being black is simply a matter of biology. It is also a matter of history, of resentments, of poverty, of food (particularly in the South), of religion, and much more. There will probably always be tension when race is being discussed, or, at least, there will be tension in the foreseeable future. The philosophers I have just discussed are all very careful to stay away from such issues as intelligence – although my suspicion is that at the popular level many people do still have strong feelings on the subject (perhaps paradoxically now putting Asians above Europeans on the intelligence scale). Interestingly, as Kitcher notes, in America there is far less of a barrier to white–Asian interbreeding than to white–black. And as we know full well, on a fairly regular basis there are other self-appointed race pundits who happily plunge in with claims about the relative, biologically based, intellectual abilities of different human groups. One wonders what these people will make of the finding that white people and Asians are walking around with Neandertal blood coursing through their veins, whereas Africans alone are racially pure!

Perhaps the best conclusion is one that applies to all of the topics discussed in this chapter. There are important and still pressing questions about human nature. Biology matters and evolution has been a crucial causal factor. Charles Darwin himself was quintessentially a child of his time. However, that is no good reason to reject the tools of inquiry that he left to us. This said, it is admittedly not always easy to know where to go from here, in terms of both teasing out the true state of affairs and knowing what to do with new knowledge. But pretending that there are no issues (or that they will go away with neglect) or that biology (perhaps because of dreadful past misuse) does not matter are not the right options. And recognizing even that is a good start.

8   From eugenics to medicine

In 1890, Alfred Russel Wallace, the co-discoverer of natural selection, wrote: "In one of my last conversations with Darwin he expressed himself very gloomily on the subject of the future of humanity, on the ground that in our modern civilization natural selection had no play, and the fittest did not survive. Those who succeed in the race for wealth are by no means the best or the most intelligent, and it is notorious that our population is more largely renewed in each generation from the lower than from the middle and upper classes" (Wallace 1900, I, 509; first published in the *Fortnightly Review* [1890]). This fits with what Darwin himself had written in the *Descent of Man*.

> With savages, the weak in body or mind are soon eliminated; and those that survive commonly exhibit a vigorous state of health. We civilised men, on the other hand, do our utmost to check the process of elimination; we build asylums for the imbecile, the maimed, and the sick; we institute poor-laws; and our medical men exert their utmost skill to save the life of every one to the last moment. There is reason to believe that vaccination has preserved thousands, who from a weak constitution would formerly have succumbed to small-pox. Thus the weak members of civilised societies propagate their kind. No one who has attended to the breeding of domestic animals will doubt that this must be highly injurious to the race of man. It is surprising how soon a want of care, or care wrongly directed, leads to the degeneration of a domestic race; but excepting in the case of man himself, hardly any one is so ignorant as to allow his worst animals to breed. (C. Darwin 1871, I, 168)

Eugenics

Darwin was responding to scares started by his cousin Francis Galton (half-cousin, actually; they shared Erasmus Darwin as a grandfather), who in his recently published *Hereditary Genius* (1869) argued that intellectual ability was something transmitted by heredity from earlier generations rather than simply a matter of environment and education, and that it would be "quite practicable to produce a highly gifted race of men by judicious marriages during several consecutive generations" (Galton 1869, 1). We see now both the positive side and the negative side to what was to be labeled the science or hope of "eugenics" – on the one hand one can try to make for a better race of humans, and on the other hand one can try to prevent the decline of the present race of humans. Either way, one is using one's knowledge of evolution and its processes to improve the state of humankind, individually or collectively. (Kevles 1985 is the definitive study of eugenics.)

Galton was an enthusiast and he passed this on to others, notably to one of the major figures in the history of statistics, the mathematician and general polymath Karl Pearson (who was British despite a passion for German culture that led him to change his name from Carl to Karl). A major handicap, however, was the lack of an adequate theory of heredity (Provine 1971; Ruse 1996). Mendelian genetics was rediscovered at the beginning of the twentieth century and at once came to Britain thanks to the passion of William Bateson (1902). With personalities counting as much if not more than pure science, Pearson allied himself with the doubters and became a leader of the opposition, the biometricians, a group who used statistics and masses of empirical data to tackle heredity purely at the observable level – what today is called the "phenotypic" level, as opposed to the "genotypic" level, the level of the genes. Things righted themselves (if that is the correct term for such a belief system) in the work of the even greater statistician and founder of population genetics, Ronald A. Fisher (1930). He was a fanatical eugenicist, and it is clear that much of his work on evolution was fueled by this (Box 1978). A very conservative Christian, Fisher believed that God had created humankind through a process of progressive evolution and that it is our task here on Earth to see that this achievement is cherished and maintained. Unfortunately, the poor and inadequate have a tendency to reproduce more than the gifted and worthy, a worry shared by virtually all of the eugenicists (and one that occupied several pages of the *Descent*). It seems that, if anything, humans – civilized humans that is – are set to decline, thanks to the ways in which they have improved their material lot and now no longer want the burden of huge families. Darwin was not alone in drawing attention to the fact that the feckless Irish were given to having many children whereas the stern and frugal Scots had only a few offspring. Darwin inclined to think that nature itself might redress the balance, for young cared-for Scots are more likely to grow to full adulthood than young neglected Irish, but Fisher felt we have a moral (and religious) obligation to redirect things ourselves. In particular, he wanted the upper classes to reproduce more – he himself had a very large family with a young wife chosen especially for her breeding potential – and went so far as to propose a system of family allowances ("baby bonuses") to be restricted precisely to those with the necessary "upper-class" qualities.

In America also one found enthusiasts for eugenics, and indeed it can be argued that this enthusiasm, along with the needs of agriculturalists, was the major motive force behind the early years of Mendelian genetics. Reflecting the place and times, much of American genetics was bound up with worries about race, not just the black–white divide, but the large influx of immigrants from eastern and Mediterranean Europe, Jews and poor Catholics whose increasing numbers in the large cities worried mightily the white, Anglo-Saxon, Protestant population. That *The Passing of the Great Race*, published in 1916, was written by a man (Madison Grant) who was first and foremost an upper-class New York socialite tells us much. On the strength of crude measures of intellectual ability, many states passed laws demanding the sterilization of the eugenically unfit. In the words of Supreme Court Justice Oliver Wendell Holmes, giving a ruling in 1927: "Three generations of imbeciles are enough." Laws were also passed bringing to an end the large influx of poor immigrants from Europe. Some did indeed recognize that in a generation the children of immigrants were performing as well as if not better than the children of those already established, but there is no stopping a prejudice whose time has come (Gould 1981).

A number of things served to dampen the passion for traditional eugenics. One was the realization, based on a better grasp of genetics, that simple formulae for overall eugenic improvement were doomed to failure. Especially if an undesirable feature is controlled by a recessive gene, it could take literally hundreds of generations to eliminate it – by which time, of course, new mutations could have brought numbers up again to the original levels. Another, and clearly a major factor, was the horror at the practices of the Nazi state in Germany in the 1930s, where eugenics was enforced rigorously through law and where many of society's unfortunates had their lives terminated in the name of racial purity and perfection. At the end of the Second World War, when it was seen fully how readily eugenics can slide into racial extermination, many (particularly in the USA, with its high number of Jewish academics and professionals) recoiled with horror at the very thought of serious human genetics, let alone proposals for improving our species through biological intervention. This tied in with what we saw in the last chapter, namely the downplaying of the significance of the notion of "race."

With few exceptions, the Nobel laureate William Shockley being one, few have wanted to return to programs for improving the race through directed breeding. However, albeit slowly, human genetics was pursued, especially in the UK through the efforts of Lionel Penrose, who held a chair at University College London – a chair that had earlier been held by Pearson and originally funded by Francis Galton. Penrose was overtly scornful of traditional eugenics, especially about links between family size and national decline. Through him and his co-workers and students, the emphasis was linked increasingly to medical concerns, and indeed some of the most important work was on heritable diseases in humans (Laxova 1998). Penrose himself was particularly interested in phenylketonuria, PKU; a condition which can now be treated through diet, but which if untreated leads to mental retardation. As is well known, work like this broadened and in the second half of the twentieth century was formalized as a distinct part of the medical scene, now relabeled under the more acceptable term of "genetic counseling." To show how very far we have come from the nightmares of the 1930s, take the case of one of the best-known genetic ailments, Tay–Sachs disease, a neurological nightmare that leads to paralysis and death, usually about the age of four. This is a disease that affects mainly certain select groups, one of which is Ashkenazi Jews, that is Jews with ancestors from eastern Europe. It is these very people, targeted above all by the Nazis for elimination, who have been at the forefront of programs for identifying couples at risk and promoting remedial measures, namely identifying fetuses that will later exhibit Tay–Sachs and then aborting. (I use the word "remedial" in a non-value sense, and will return to this point later.)

Evolutionary medicine

With the coming and completion of the Human Genome Project, the mapping of the entirety of the human gene complex, and now increasingly ferreting out differences between individuals, genetic counseling is clearly here to stay. Whether we should give thanks to the original eugenicists for lighting the way or whether we should give thanks for the demise of original eugenics is a question we can leave to the historians. Let us rather now broaden our gaze, for genetic counseling is but one part of (or a parallel movement to) a much more ambitious would-be entry into the field of healthcare, namely "evolutionary medicine," the project of putting the whole of our understanding of and behavior toward those in need of attention on a firm evolutionary basis. Very much the brainchild of two men, George C. Williams, a leading twentieth-century evolutionist, whom we have encountered before, and Randolph Nesse, a University of Michigan-based psychiatrist, evolutionary medicine aims to revolutionize the field (Nesse and Williams 1994, 1995). As Theodosius Dobzhansky (1973) used to say, "nothing in biology makes sense except in the light of evolution," so Williams and Nesse would add "and nothing in medicine, either, makes sense except in the light of evolution."

Even though by 1959 (a convenient date, being the centenary of the *Origin*) neo-Darwinism was an up-and-running paradigm, it was many years before evolutionary theory was properly integrated into the biology undergraduate curriculum (and there are places even now where this is barely the case). In a way, Dobzhansky's statement was less a proud affirmation and more a plea for recognition. So likewise there has not exactly been a rush by healthcare professionals (and more pertinently their teachers) to embrace the proffered insights of evolutionary theory. However, rather than bewailing (or celebrating) this fact, let us move directly to consider the claims made in a major recent textbook, co-authored by the eminent New Zealand scientist and physician Sir Peter Gluckman. Let us follow him through an eightfold classification of the ways in which evolution can impinge on the question of disease and our risk of suffering (Gluckman *et al.* 2009, 257–76). I will leave any philosophical reflections until we have discussed this classification.

First, there is the fact that we (some individuals that is) might be in a situation for which evolution has not prepared us. Our environment or our culture has outstripped our biology. Among the items mentioned by Gluckman and his co-authors, where biology and culture get out of focus, is our inability to synthesize vitamin C, something that led to scurvy on board ships, until navies realized that daily drinks of citrus juice could avert the problem. There is also obesity, perhaps not itself an illness but obviously one leading to such. It could be caused by the built-in desire to gorge when possible, something perhaps of great value in the Pleistocene but obviously much less so in modern society. A possible third example is the oft-mentioned lactose intolerance. Since the coming of agriculture, about 8,000 years ago, there has been intense selective pressure on agriculturalists toward lactose tolerance, that is to the ability through life to tolerate milk (and milk solids) from our domestic animals. Most Europeans now have this acquired ability, but many in other parts of the world, without histories of agriculture, do not. This can cause severe discomfort. Most interestingly, a recent study suggests that Charles Darwin's forty-year adult sickness may have been rooted in lactose intolerance, something rare but not unknown among Europeans. We know from his wife's cookbook that the Darwins ate a typically upper-middle-class diet, very rich in dairy products. Darwin's health often improved temporarily when he went off to health spas. It could well be that it was the meager diet at such places that was really doing him good, rather than the somewhat strenuous "cures" – like being plunged into freezing-cold water and then vigorously toweled – that he undertook, at great expense.

Second on the list of evolutionary factors affecting health and sickness are life-history-associated elements. Most obvious here are the ailments of old age. Natural selection cares about getting us up to prime breeding condition and then keeping us fit as long as we are actively involved in child care and rearing. After that, we are on our own, and a very lonely "on our own" it can be, too. We are much less able to handle infections and traumas and also have all of the diseases of old age – diseases that we would have evaded (because we would already be dead) in earlier times. Sometimes there is a direct connection between things that are useful earlier in life but harmful later on. Stem cells in tissues are a good example. While we are growing and reproducing, stem cells in tissues are of value because they promote tissue maintenance and repair. Unfortunately, later in life, they can lead to neoplasia, the abnormal proliferation of cells, which may perhaps be malignant.

Third are excessive and uncontrolled defense mechanisms. Things like coughing, vomiting, and diarrhea are unpleasant in themselves, but they are fairly obvious ways in which a body tries to expel or ward off invading organisms. This can backfire if the mechanisms go into overdrive – dehydration following severe gastroenteritis is a case in point. Knowledge of the evolutionary significance of the mechanisms can be important in treatment. It is a commonplace that when we are sick we are often (usually) lethargic, disinclined to do anything very strenuous. There is evidence that this lethargy is part of our biology, slowing us down so the body can concentrate on fighting the sickness. Exercise when sick can be counterproductive. Fevers often fall into the same category. The usual advice is to take two aspirin and go to bed. But fevers are thought to be significant in fighting infection. Perhaps less so because they kill off bacteria directly than because they initiate the production of certain proteins ("heat shock proteins") that can circulate in the blood and that have powerful anti-bacterial functions.

Fourth comes "losing the evolutionary arms race against other species." This is a very well-known and -attested phenomenon. As soon as some new drug combating infection comes on the market, the organism or organisms against which it is directed are under huge selective pressure to develop ways to respond. Given the high rate at which organisms like bacteria reproduce – and their sheer number – it is hardly surprising that it is but a few years, if that, before resistant strains are known. Figure 8.1 tells all. Look for instance at penicillin, introduced around 1942 and invaluable during the Second World War. Within a year of the war's end, resistant strains were emerging. What makes this all a major problem is that nowhere do such resistant organisms emerge more quickly and strongly than in hospitals, the very places where people need most to be protected. This stems from a number of reasons, including the number of people who are sick, the high use of antibiotics, the ways in which staff can transmit diseases, and more. Expectedly, the fight against disease is complicated by the fact that resistant organisms resist in many different fashions. There is no one way in which they evade drugs, a way that could be explored in the hope of finding a one-step solution for everything.

Fifth is the matter of design or evolutionary constraints. We have encountered this notion already in the more general context. Some things highlighted here by Gluckman are perhaps less constraints but rather things left over by the evolutionary process. The appendix is a case in point. For

Antibiotic deployment

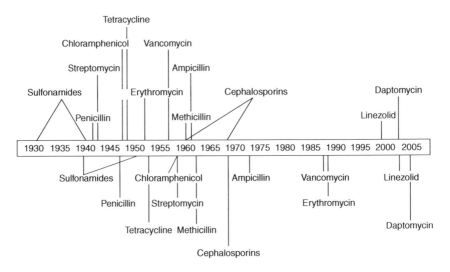

Antibiotic resistance observed

Figure 8.1 Timeline of antibiotic deployment and the development of antibiotic resistance. Note the rapid appearance of resistance after deployment (from Clatworthy *et al.* 2007).

us, appendicitis can be fatal and yet the appendix probably has little or no function. It is a throwback to the days when our ancestors ate huge and probably near-exclusive amounts of herbage and needed the appendix for digestion. (Recently it has been suggested that the appendix might have a role in the immune system. See Martin 1999.) More obviously a constraint bringing about compromise is that determining the size of the human brain at birth. The larger the brain the sooner the child will grow up to full size. However, the larger the brain the more danger to the mother, who has a birth canal determined in size by the demands on the pelvis for upright walking. Another constraint is the lower back, which has loads and stresses upon it thanks to upright walking, quite beyond anything experienced by the apes. And as another example, it may be that male breast cancer (about 1 percent of all cases) comes about simply because there is just no easy way of getting rid of the genes which, when properly primed, cause functioning breasts in females.

Sixth is disease due to the direct effects of natural selection as it "balances" good effects against bad. The classic case here is that of sickle cell anemia. In parts of Africa, malaria is a dreadful threat, killing sizable proportions of the population. Any gene, therefore, that offers protection against malaria is going to be under strong positive-selection pressure. It turns out that there is a gene that offers protection, but the catch is that it offers such protection only if it is present in a single dose – more formally, if it is heterozygous to the normal or wild-type allele (gene). It affects red blood cells in such a way that if the body is invaded by malaria then infected red blood cells collapse and are removed by the white blood cells. Unfortunately, two doses (the sickling allele is homozygous) cause the red blood cells to collapse into a tell-tale crescent or sickle shape, and the carrier generally dies young of anemia. It is a very simple piece of mathematics to show that in a population the devastating ill-effects of two sickle cell alleles are balanced by the good effects (the malarial protection) of one sickle cell allele – and this persists for generation after generation, unless something external disrupts things. Another possible case of "balanced heterozygote fitness" might involve cystic fibrosis. It is thought that it may be caused by genes that in lesser amounts protect against typhoid or tuberculosis.

Seventh we have sexual selection and its effects. There are good reasons why fighter pilots tend not to be older men and women of sixty. Thanks to sexual selection, it is young males who are more prepared to take risks and to do dangerous things. It is they who have the right hormones pumping through their bodies, making them ready to fight and to compete, directly or indirectly, for females. Of course, young male humans are not just sexual aggressors. We have all been selected for sociality, the ability to live in groups. This requires moderation and tempering other desires. So at the very least we may have psychological conflicts and at worse violence and the injuries to which this and other forms of risk-taking can lead. Whether, as seems fashionable today for film stars and sports celebrities, extreme sexual desire and behavior should be labeled a sickness is an exercise left to the reader.

Finally, eighth, come the "outcomes of demographic history." We have already touched on these issues earlier in this chapter and before. Gluckman and his fellow authors are referring to the asymmetries that we find in groups with respect to various genetically caused illnesses as a result of evolutionary history. Ashkenazi Jews and Tay–Sachs disease is a case in point. There is nothing special about being Jewish that makes for susceptibility to Tay–Sachs disease. It is rather that by an accident of history the mutation got into the population, and because breeding (up to now, at least) has tended not to occur across population borders, Ashkenazi Jews are especially susceptible. In this case, it is primarily social factors that have set up the barriers. In fact, these barriers have been to a great extent dismantled in the USA, where increasingly there are unions between Jews and Gentiles. In other cases, the barriers have been more physical or geographical. This may be combined with bottlenecks when population numbers were greatly reduced before expanding again. The inhabitants of Finland may exhibit all of these things. They came in small numbers across the Baltic from southern Europe and once settled were isolated by geography and climate. Expectedly, they show patterns of illness that distinguish them from others. For instance, there is a comparatively low incidence of Huntington's chorea, of cystic fibrosis, and of PKU. There is a high incidence of type 2 diabetes and cardiovascular disease. None of this is to exclude the possibility of environmental factors. Finland is very different from Italy in both winter and summer. It is to say that evolutionary biology may have been very important.

Presuppositions

We move on now to ask about matters of possible philosophical interest, some of which are immediately obvious. We all know that abortion and sterilization, especially enforced sterilization, are highly contested moral issues. With the Catholic Church taking strong stands against both, there is little surprise that it was one of the chief opponents of eugenics, especially of the negative kind which wanted to restrict the reproduction and the production of the unfit. One suspects that with the growth in influence of the Catholic Church in the USA (not the least on the Supreme Court), any eugenic proposals like those floated (and often enacted) in the USA in the early twentieth century would have a much rougher ride today. Clearly, then, there are philosophical issues here; perhaps general issues posed, possibly exacerbated, by evolutionary medicine (broadly conceived to include such things as genetic counseling) rather than ones uniquely formulated by evolutionary medicine.

It is obvious also that there are going to be important conceptual and epistemological issues at stake. As one might expect of a field that was kick-started by George C. Williams, today's evolutionary medicine tends to be hard-line Darwinian, and individual selectionist at that. This does not mean that everything is thought to be an adaptation. No Darwinian, certainly not Charles Darwin, has ever made that claim. We saw earlier that evolutionary medicine supposes that there are all sorts of places where things can get out of adaptive focus. There may be a lag between what was adaptive in the past and what is adaptive now. What is adaptive for one organism (a parasite) may not be particularly adaptive for another organism (us). Constraints and compromises and (what Darwin called) vestigial organs are another set of places where adaptation does not rule untroubled. Then there is the fact that sexual selection might be adaptive one way, but clearly might be highly counter-adaptive another way. And the final item, where we look at the effects of history on groups, shows that random factors – the founder effect, especially – may play a crucial role in human health and disease. All of this, it goes without saying, is pretty standard Darwinian theory and has been stressed again and again in the century and a half since the *Origin*. The important point is that no new theory is demanded by the entry of medicine into the world.

None of this is to deny that the touchstone, the expected norm, is adaptive advantage. The whole point of evolutionary medicine is that we are looking at the body as a product of natural selection, and we expect to see adaptive advantage. In many cases we see this straightaway. Eyes are for seeing and blindness is an affliction. Noses are for smelling, although as Gluckman points out our evolutionary history rather points to the obvious fact that we rely a lot less on smell than do other mammals like dogs and hence the sense of smell is nothing like as crucial as the sense of sight. We have special schools for children who are blind. We have special schools for children who are deaf. We do not have special schools for children who lack a sense of smell. Perhaps in the Pleistocene we might have needed them, because then a sense of smell might have been more vital – sniffing out meat that had gone off, for instance.

The commitment to individual selection is absolutely crucial in some instances where it is thought that evolutionary medicine has made triumphant breakthroughs in understanding. Harvard biologist David Haig (2008), for instance, has studied the relationships between mothers and offspring. You might think that here at least we are going to get one big, happy family, but Haig – drawing on earlier individual-selectionist thinking by one of the founders of sociobiology, Robert Trivers (1974) – notes that we may well get "parent–offspring conflict." What is in the interests of a mother may not be in the interests of a child and vice versa. Really, this is obvious as soon as you think about it. If a mother has two children it may be in her interests to split her attention between them or perhaps to give the younger child more attention. It does not follow that this is in the biological interests of the children, particularly the older child, even after we have factored in the relationship between the children (especially if there are different fathers). Haig applies this thinking to the circulation of the mother's blood. It is in the interest of the fetus to raise the proportion that goes to it; it is in the interest of the mother to moderate the circulation of the blood that goes to the fetus. The way this is played out is by the resistance that can be set up to the mother's circulation, and that is a function of her blood pressure. Blood pressure drops early in pregnancy, and Haig's claim is that this represents the triumph of mother over fetus. Then later it starts to rise as the fetus now directs more blood in its direction – from virtually nothing at the beginning of the pregnancy to 16 percent at the end of the pregnancy.

Of course in a way you might rightly argue that this all functions in such a way that the fetus does well but the mother survives – they have those interests in common. But sometimes in pregnancy women develop preeclampsia, a very dangerous condition associated with very high blood pressure, together with lots of protein in the urine. The obvious interpretation is simply that something has gone wrong. Haig suggests rather that it might be a move of desperation on the part of the fetus. If for some reason it is not getting sufficient nutrients, it is in its interests to up the mother's resistance to that blood flow benefiting her (the mother) alone, and the way to do this is by taking the blood pressure up even to dangerous levels. Interestingly, preeclampsia is more common with twin pregnancies, and this of course is precisely a case where one individual fetus might not be getting enough blood. The fetus is gambling that it might be better off taking what it can now, even though it runs the risk of losing out on care later. The mother's interests do not enter into the equation, or only secondarily. About as individual-selectionist a perspective as you could get.

The point is not that we now have a solution to preeclampsia or that we should refrain from interfering because this is "nature's way," or some such thing. At most, we may now have some true insights: knowledge is the beginning of successful action. Nor is the point that individual selection theory is right and all-conquering. Rather that this seems to be the general pattern in evolutionary medical explanations today and needs to be recognized by those who would propose alternative explanations. Clearly there is need of conceptual analysis, for already some working in the field have appropriated terms like "multi-level selection." A case in point comes in a recent discussion (by Carl T. Bergstrom and Michael Feldgarden) of ways in which one might apply insights from evolution to the creation of new barriers to invasive organisms. They point out that the dangers posed by bacteria are often not from the individual bacteria as such, but rather when they are in groups and start acting together. In other words, when we have a "quorum." Perhaps, therefore, a solution might be to trick the bacteria into thinking (highly metaphorical language here!) that there is no such quorum. Moreover, and very desirably, when the bacterial social behavior is disrupted, it might not rebound as quickly as one might suppose. They write (referring to the ideas of others):

> Where bacterial cooperation occurs, it is not an unavoidable consequence of direct individual selection as antibiotic resistance usually is, but rather a finely balanced consequence of multilevel selection. Thus if bacterial cooperation is disrupted, it may not return as readily as individually selected traits. To see how this might work, imagine a population of bacteria in which social behavior has been halted by disrupting quorum sensing. Whereas with conventional antibiotics the first antibiotic-resistant mutant has a substantial growth advantage, with quorum-sensing disrupters the first resistant mutant has a growth disadvantage. It provides a public good by producing constitutively, but it receives no benefits from the other members of the population who are not producing due to the quorum sensing disrupter. Moreover, because these behaviors are selected at the population level, if resistance does evolve it is likely to do so on the time scale of populations, rather than on the time scale of individuals. While a bacterium may reproduce in a matter of hours, populations often turn over on scales of weeks to months and thus resistance to quorum sensing disruptors is likely to evolve much more slowly than does resistance to conventional antibiotics. (Bergstrom and Feldgarden 2008, 134–35)

Reading this the first time leaves a clear impression that group selection is at work – "behaviors are selected at a population level." However, if you look carefully, no such mechanism is really being proposed. The behaviors occur at the population level, but because they do not at first serve the interests of individuals – "it receives no benefits from the other members of the population" – they do not spread quickly. Indeed, one might even ask why they spread at all. "Multilevel selection" is a term being used not to incorporate group selection, but to acknowledge that individual selection can have group effects that are going to be important to the individual.

Sickness and health

Apart from these issues to do with the kind of evolutionary theory being proposed or rather presupposed by those working in the field of evolutionary medicine, there are clearly other topics of great philosophical interest. For instance, very obviously there are issues to do with testing and how one might check out theories that apply to human beings. One cannot simply run experiments as one might on mice or rabbits. However, I want now to turn to a topic that has already been touched on in earlier chapters (notably during the discussion of sexual orientation) but that now merits more explicit attention, for it is something that lies behind any philosophical inquiry into the nature of medicine. I refer to the concepts of sickness, disease, and health. In recent years, much has been written on these topics, and it is important to see how they play out in the context of evolutionary medicine. This discussion will be a two-way process. What have the philosophers' discussions to say about evolutionary medicine and what has evolutionary medicine to say about the philosophers' discussions?

Leaving for a moment the question of health (and whether it is just a reverse of sickness), there are two basic approaches to the key problem of disease. These usually go under the headings of "naturalism" and "normativism," although other terms have been proposed (Ruse 1988b). Philip Kitcher, for instance, proposes "objectivism" and "subjectivism."

> Some scholars, objectivists about disease, think that there are facts about the human body on which the notion of disease is founded, and that those with a clear grasp of those facts would have no trouble drawing lines, even in the challenging cases. Their opponents, constructivists about disease, maintain that this is an illusion, that the disputed cases reveal how the values of different social groups conflict, rather than exposing

any ignorance of facts, and that agreement is sometimes even produced because of universal acceptance of a system of values. (Kitcher 1997, 208–209)

Whatever the language, you can see that the key divide is between those who think that disease is something "out there," that can be defined in terms of actual physical facts, and those who think that disease is necessarily a value notion, and as such is a matter of subjective or cultural ideas or themes or preferences.

Start with the naturalist position. The standard account is provided by Christopher Boorse (1975; also 1977, 1987). He states flatly that: "On our view, disease judgments are value neutral ... their recognition is a matter of natural science, not evaluative decision" (Boorse 1977, 543). But how does one cash out the reference to natural science? In some sense, it has to be a matter of what is normal or natural for the species. "There is a definite standard of normality inherent in the structure and effective functioning of each species or organism ... Human beings are to be considered normal if they possess the full number of ... capacities natural to the human race, and if these ... are so balanced and inter-related that they function together effectively and harmoniously" (554). But how are we to articulate the "definite standard of normality"? There's the rub! Suppose you just work statistically, and argue that the standard is the majority is the norm. Does this mean that being a minority in itself makes you sick? Apart from the tricky issue of sexual orientation – is a homosexual sick simply because he or she is in the minority? – think of the case of sickle cell anemia. We certainly want to say that if the notion of disease comes in anywhere, it comes in here. But it is far from obvious that we are making this judgment simply because the sufferers are in the minority. We are making it because they are in desperate pain and will die young.

Perhaps, therefore, we should think more in terms of effective and harmonious functioning. Ignoring the group-selection hints in the above definition, presumably what we are now thinking of is to be cashed out in evolutionary terms, that is to say survival and reproduction. In a way, that seems to be pretty good, and attractive from the viewpoint of evolutionary medicine. If someone has a childhood leukemia, they are diseased because their prospects of survival and reproduction are much reduced. Similarly, if someone is losing out in an arms race with bacteria, again prospects of survival and reproduction are grim. However, we obviously run into problems very quickly. Suppose someone is vomiting and has diarrhea and a high temperature. Evolutionary medicine says that this is the body's way of kicking in and combating an infection. Do we want to say that such a person is not sick? "Pull your socks up and don't whine!" Sickle cell anemia makes the situation even worse. Here we have something positively promoted by natural selection, at least in the sense that selection keeps the numbers up so that the heterozygotes do better in the struggle for survival and reproduction than they would otherwise. It is a very typological view of species – one that the late Ernst Mayr (1942) spent a very long lifetime combating – to say that a minority, being produced by natural selection, is not in some way typical of the species and even in some way part of natural functioning. As Randolph Nesse is always saying, you have got to stop thinking of natural selection as promoting health and happiness (whatever these might be). It is in the survival and reproduction business completely and utterly (Nesse and Williams 1994).

Proximate versus final causes

Perhaps at this point it would be useful to invoke the distinction between proximate and ultimate causes. An evolutionary perspective obviously focuses on ultimate causes, or final causes in the traditional language. Perhaps in medicine we should be looking always or primarily at proximate causes.

> Schaffner (1993) has argued very convincingly that although medicine might use teleological talk in its attempts to develop a mechanistic picture of how humans work, the teleology is just heuristic. It can be completely dispensed with when the mechanistic explanation of a given organ or process is complete. Schaffner argues that as we learn more about the causal role a structure plays in the overall functioning of the organism, the need for teleological talk of any kind drops out and is superseded by the vocabulary of mechanistic explanation, and that evolutionary functional ascriptions are merely heuristic; they focus our attention on "entities that satisfy the secondary [i.e. mechanistic] sense of function and that it is important for us to know more about" (1993, 390).
> (D. Murphy 2008)

Prima facie, this is an attractive move to make. A person with a temperature and the trots is sick no matter what the reasons in the long run. Look at what is making life so very difficult right now and get on with the process of helping and healing. Likewise the child with an appalling anemia is sick, has a disease, no matter what the ultimate reasons for this. That its siblings are thriving is in a way irrelevant. We want to know what causes the anemia, meaning the proximate causes, and how to tackle it.

Note that this is a broad-ranging conceptual argument. You might load it down with additional points, for instance about the possibility that a disease has no direct final cause. It is not an adaptation, but a failure in adaptation. Or it was never really an adaptation in the first place, but perhaps a byproduct or the result of a constraint – the sort of thing that Stephen Jay Gould called a spandrel (Gould and Lewontin 1979). However, we saw earlier that the evolutionary medicine supporter has the resources to deal with these issues, because the form of Darwinism that is presupposed takes these issues into account. (Whether practitioners always take them as fully into account as they should is perhaps another matter.) The question is whether looking at final causes, thinking teleologically, is a mistake in the first place. At best it can be used as a tool for discovery, as a heuristic. Here the response of the evolutionary medicine supporter will be that, however you want to define terms like disease and health, if you are in the business of healthcare then you really must look at final causes, you must ask Darwinian questions about adaptation. These are not merely heuristic. They are fundamental, and crucial to formulating adequate treatment. Should you give someone a couple of aspirin to bring down the temperature, or should you tell them to take it easy and tough it out? What about cases of preeclampsia? Should we just be focusing on the mother, or should we recognize that this may be a cry for help from the fetus and in treating the situation try to see that the fetus's needs are also taken into account? Should we see that evolution might tell us a lot about the fetus's needs? When faced with a difficult childbirth, should we recognize that "natural" may not necessarily be the best thing? We are faced with a compromise, and natural selection has not been able to perform miracles. Hence, interventions in the form of cesarean births or at the least episiotomies are not to be proscribed in the cause of some false beliefs about naturalness or some such thing.

What about the actual use of language? Notwithstanding the significance of finding the evolutionary causes behind the phenomenon, it is hard to see how under any circumstances one would not want to speak of sickle cell anemia as a disease. But perhaps in other cases one would want to modify the language. Perhaps a lot depends here on how revisionist one is going be about language use, or whether one is going to be conservative about these things. This is not a totally insignificant matter or a mere question of taste. For instance, the medical definition of "obesity" today encompasses what in the past might have been described as "pleasingly plump," clearly a move made with the hope that those who are overweight, even if not grossly so, will be shocked into taking some remedial measures. In the same vein, namely of improving healthcare through the revision of language, one could possibly see a case for distinguishing cases where natural selection is working for the evolutionary ends of the individual (no matter how unpleasant) as opposed to cases where natural selection is working for the ends of others (healthy siblings, healthy babies) and cases where natural selection is simply failing (losing out in an arms race). Perhaps already, assuaging some of the worries of the linguistic conservative, we do some of this. Knowing the true state of affairs, you might want to say that the disease is the bacterial infection that your body is fighting. Having a high temperature is an unpleasant side-effect but certainly not a disease in itself or even part of the disease proper, whereas perhaps a swollen organ brought on by the bacteria is part of the disease proper.

Values

All of this discussion is clearly a bit truncated and distorted because we are not bringing in something that even the naturalist must take account of in some way, namely values. Why do we want to say that sickle cell anemia is a disease? Ultimately, because it is clearly unpleasant. People with sickle cell anemia hurt. At some level this is undeniably what lies behind Schaffner's urging us to think in terms of proximate causes. It is at the proximate level that pain and suffering occur. Final causes may be useful to understanding, but basically by definition they are not dealing with the here and now, and it is this that ultimately counts in medicine. As it happens, Boorse himself acknowledges this issue, for he makes a distinction between disease and sickness or being ill. It is the second term that carries the burden (if such it be) of values. Being ill is having a disease that we do not want, because it is unpleasant. Notice that values alone will not do the trick. My body must be broken down in some way biologically. I am in jail, awaiting execution for a crime that I did not commit. Undoubtedly I will be very sad, and a lengthy spell on death row might make me clinically mad. But my essential sadness is no illness but a natural reaction to misfortune. (I specify that I did not commit the crime, because in the case of many of those who did commit crimes their mental health is already in question.)

Of course, introducing values does not solve the epistemological problems about defining disease purely naturally. Perhaps, therefore, the time has come to move right over to a normativist account of disease, what Kitcher calls constructivist. H. Tristram Engelhardt Jr. (1976) is the point person here: "We identify illnesses by virtue of our experience of them as physically or psychologically disagreeable, distasteful, unpleasant, deforming" (259). Obviously, this is not enough. I am not ill if I am unjustly condemned to death. At once the normativists start moving over toward the naturalist side. We identify them "by virtue of some form of suffering or pathos due to the malfunctioning of our bodies or our minds. We identify disease states as constellations of observables related through a disease explanation of a state of being ill" (259). Note that Engelhardt does seem to be assuming some kind of evolutionary, teleological understanding of causation. Others in the normativist camp seem more wedded to just proximate causes. Reznek (1987), for instance, wants to get away from the notion of malfunctioning and talk more in terms of abnormal processes. A major point here is that, once you give up the supposedly objective science as your measure of disease and move to something like abnormality, not only are you moving to proximate causes but you are moving into the realm of culture, where what counts as an abnormality itself requires a value judgment. In other words, medical problems are what medical people deal with! A little less tautologically, what is to count as a disease requires a value judgment in itself. For instance, in many societies the desire of a man to have sex with as many women as possible is considered normal if sometimes socially awkward; in the USA, apparently, it is an ailment calling for treatment.

The exponent of evolutionary medicine clearly cannot accept this at all. There can be agreement that pain and suffering are important in judging whether someone is sick, and that this might go back to the question of whether someone has a disease. There might be some sympathy for defining diseases in proximate-cause terms, so long as this in no way denies or leads away from the essential importance of thinking in terms of adaptation, of final causes. But ultimately whether there is something medically wrong cannot be a judgment from culture. Culture might be important in the judgment. The first of Gluckman's categories deals with ailments that come from rapid changes in environment, and culture is a key factor here. But the judgment is not itself cultural. In the end, it all comes down to survival and reproduction. Some do, some don't. It is as simple (or complex) as that.

Health

What about the flip side, what about health? To invoke Nesse again, natural selection does not care about you being healthy. It cares about survival and reproduction. Whether it is a disease or not, suppose your sexual obsessions are making you downright miserable. Instead of being able to settle down to a comfortable evening of reading the *Critique of Pure Reason*, you feel compelled once again to haunt the singles bars, engage in trivial and insincere conversation, all for the chance of a night of sex. If this is a better way of passing on your genes, then so be it. Unless it can be shown that your behavior backfires in some way, perhaps through the spread of STDs or perhaps because those children who are born because of your behavior do not receive proper parental care, it really doesn't seem that evolutionary medicine has a dog in this fight.

Obviously this is a little bit extreme. Are there more subtle ways in which biology might be connected to health? Some people want to define health in a way that refers to oneself essentially. Others, like the German philosopher Hans Gadamer (1996), prefer to put things in a social context: "it is a condition of being involved, of being in the world, of being together with one's fellow human beings, of active and rewarding engagement in one's everyday tasks" (113). And some want to combine the two. The World Health Organization defines health as "a state of complete physical, mental and social well-being and not merely the absence of disease or infirmity." (This is taken from the constitution of the World Health Organization, a body of the United Nations. It was formulated in 1946 and became active in 1948.) However one decides, as the WHO definition makes clear, one is probably not going to define health purely in terms of survival and reproduction. Having a sense of fulfillment and being worthwhile is part of being healthy, and obsession with numbers of children is surely odd to the point of imbalance somewhere. Nevertheless, having children may be a very important part of what one considers full and healthy living. There are those who regard DINKS by choice (Double income, no kids) as if not sick then sadly truncated as human beings. Moreover, unless you are essentially free from disease and handicap you are probably less likely to have total fulfillment and thus are less likely to be judged totally healthy. So biology surely does come in somewhere, and the pertinence of the evolutionary approach is not to be denied totally or even in large part.

Exactly how this is all to be worked out is without doubt a task for the future, and one suspects that philosophers could have much of worth to contribute. And this reflection, put in a broader context, is a good point on which to end our discussion. For all of its historical antecedents, evolutionary medicine as a formal approach is a relatively new discipline, perhaps with much promise but with far to go, both as science and medicine and as something that makes part of medical organization and (very importantly) teaching. It raises some philosophical issues of great interest, and those trained in the field both could and should get involved. But by this stage of our inquiry, this is hardly a surprise, for as we have seen throughout this book, the whole field of human evolution is one that is pregnant with philosophical questions of great interest. My hope is that this short introduction will stimulate others to pick up the torch and to carry on the inquiry.

Bibliography

Alexander, R. D. 1974. The evolution of social behaviour. *Annual Review of Ecology and Systematics* 5: 325–83.

Allen, G. E. 1978. *Thomas Hunt Morgan: The Man and His Science*. Princeton University Press.

Andreasen, R. O. 1998. A new perspective on the race debate. *British Journal for the Philosophy of Science* 49: 199–225.

   2000. Race: biological reality or social construct? *Philosophy of Science* 67: S653–S666.

Aquinas, St. Thomas. 1975. *Summa Contra Gentiles*, trans. V. J. Bourke. University of Notre Dame Press.

Arensburg, B., A. M. Tillier, B. Vandermeersch, H. Duhay, L. A. Schepartz, and Y. Rak. 1989. A Middle Paleolithic human hyoid bone. *Nature* 338: 758–60.

Atran, S. 2004. *In Gods We Trust: The Evolutionary Landscape of Religion*. New York: Oxford University Press.

Ayala, F. J. 1988. Can "progress" be defined as a biological concept? In *Evolutionary Progress*, ed. M. Nitecki, 75–96. University of Chicago Press.

   2009. Molecular evolution. In *Evolution: The First Four Billion Years*, ed. M. Ruse and J. Travis, 132–51. Cambridge, Mass.: Harvard University Press.

Bakker, R. T. 1983. The deer flees, the wolf pursues: incongruencies in predator-prey coevolution. In *Coevolution*, ed. D. J. Futuyma and M. Slatkin. Sunderland, Mass.: Sinauer.

Barrett, P. H., P. J. Gautrey, S. Herbert, D. Kohn, and S. Smith, eds. 1987. *Charles Darwin's Notebooks, 1836–1844*. Ithaca, NY: Cornell University Press.

Bates, H. W. 1862. Contributions to an insect fauna of the Amazon Valley. *Transactions of the Linnean Society of London* 23: 495–515.

Bateson, W. 1902. *Mendel's Principles of Heredity. A Defence, with a Translation of Mendel's Original Papers on Hybridisation*. Cambridge University Press.

Beatty, J. 1981. What's wrong with the received view of evolutionary theory? *PSA 1980* 2, 397–426.

2001. Hannah Arendt and Karl Popper: Darwinism, historical determinism, and totalitarianism. In *Thinking about Evolution: Historical, Philosophical, and Political Perspectives*, ed. R. S. Singh, C. B. Krimbas, D. B. Paul, and J. Beatty, 62–76. Cambridge University Press.

Bell, A. P., and M. S. Weinberg. 1978. *Homosexualities – A Study of Diversity among Men and Women*. New York: Simon & Schuster.

Bell, A. P., M. S. Weinberg, and S. K. Hammersmith. 1981. *Sexual Preference: Its Development in Men and Women*. Bloomington: Indiana University Press.

Bergson, H. 1907. *L'évolution créatrice*. Paris: Alcan.

Bergstrom, C. T., and M. Feldgarden. 2008. The ecology and evolution of antibiotic-resistant bacteria. In *Evolution in Health and Disease*, 2nd edn., ed. S. C. Sterns and J. C. Koella, 125–37. Oxford University Press.

Blanchard R. 1997. Birth order and sibling sex ratio in homosexual and heterosexual males and females. *Annual Review of Sex Research* 8: 27–67.

Bloch, J. I., and D. M. Boyer. 2002. Grasping primate origins. *Science* 298: 1606–10.

Boorse, C. 1975. On the distinction between disease and illness. *Philosophy and Public Affairs* 5: 49–68.

1977. Health as a theoretical concept. *Philosophy of Science* 44: 542–73.

1987. Concepts of health. In *Health Care Ethics*, 359–93. Philadelphia: Temple University Press.

Boswell, J. 1980. *Christianity, Social Tolerance, and Homosexuality*. University of Chicago Press.

Bowler, P. J. 1996. *Life's Splendid Drama*. University of Chicago Press.

Box, J. F. 1978. *R. A. Fisher: The Life of a Scientist*. New York: Wiley.

Boyd, R., and P. J. Richerson. 1985. *Culture and the Evolutionary Process*. University of Chicago Press.

Boyer, P. 2002. *Religion Explained: The Evolutionary Origins of Religious Thought*. New York: Basic Books.

Braithwaite, R. 1953. *Scientific Explanation*. Cambridge University Press.

Brandon, R. N., and R. M. Burian, eds. 1984. *Genes, Organisms, Populations: Controversies over the Units of Selection*. Cambridge, Mass.: MIT Press.

Brown, P., T. Morwood, M. J. Soejono, *et al.* 2004. A new small-bodied hominin from the Late Pleistocene of Flores, Indonesia. *Nature* 431: 1055.

Browne, J. 1995. *Charles Darwin: Voyaging. Volume I of a Biography*. New York: Knopf.

2002. *Charles Darwin: The Power of Place. Volume II of a Biography*. New York: Knopf.

Buller, D. J. 2005. *Adapting Minds: Evolutionary Psychology and the Persistent Quest for Human Nature*. Cambridge, Mass: MIT Press.

2009. Evolutionary psychology. In *Evolution: The First Four Billion Years*, ed. M. Ruse and J. Travis, 557–60. Cambridge, Mass.: Harvard University Press.

Burchfield, J. D. 1975. *Lord Kelvin and the Age of the Earth*. New York: Science History Publications.

Byars, S. G., D. Ewbank, D. R. Govindaraju, and S. C. Stearns. 2010. Natural selection in a contemporary human population. *Proceedings of the National Academy of Sciences* 107(1): 1787–92.

Campbell, B. 1972. *Sexual Selection and the Descent of Man*. Chicago: Aldine.

Camperio C. A., F. Corna, and C. Capiluppi. 2004. Evidence for maternally inherited factors favoring male homosexuality and promoting female fecundity. *Proceedings of the Royal Society of London, Series B: Biological Sciences* 271: 2217–21.

Carroll, S. B., J. K. Grenier, and S. D. Weatherbee. 2001. Homeotic genes and the evolution of arthropods and chordates. *Nature* 376: 479–85.

Cartmill, M. 1974. Rethinking primate origins. *Science* 184(4135): 436–43.

Cavalli-Sforza L. L., and M. W. Feldman. 1981. *Cultural Transmission and Evolution*. Princeton University Press.

Ceci, S. J., and W. M. Williams. 2009. *The Mathematics of Sex: How Biology and Society Conspire to Limit Talented Women and Girls*. New York: Oxford University Press.

Chambers, R. 1844. *Vestiges of the Natural History of Creation*. London: Churchill.

1846. *Vestiges of the Natural History of Creation*, 5th edn. London: Churchill.

Chomsky, N. 1957. *Syntactic Structures*. The Hague: Mouton.

Clatworthy, A., E. Pierson, and D. T. Hung. 2007. Targeting virulence: a new paradigm for antimicrobial therapy. *Nature Chemical Biology* 3: 541–48.

Clutton-Brock, T. H. 2007. Sexual selection in males and females. *Science* 318: 1882–85.

Clutton-Brock, T. H., F. E. Guinness, and S. D. Albon. 1982. *Red Deer: Behaviour and Ecology of the Two Sexes*. University of Chicago Press.

Colbert, E. H. 1992. *William Diller Matthew, Paleontologist: The Splendid Drama Observed*. New York: Columbia University Press.

Coleman, W. 1964. *Georges Cuvier, Zoologist. A Study in the History of Evolution Theory*. Cambridge, Mass.: Harvard University Press.

Conway Morris, S. 2003. *Life's Solution: Inevitable Humans in a Lonely Universe*. Cambridge University Press.

Cosmides, L. 1989. The logic of social exchange: has natural selection shaped how humans reason? Studies with the Wason selection task. *Cognition* 31(3): 187–276.

Cosmides, L., and J. Tooby. 1992. Cognitive adaptations for social exchange. In *The Adapted Mind: Evolutionary Psychology and the Generation of Culture*, ed. J. H. Barkow, L. Cosmides, and J. Tooby. New York: Oxford University Press.

1996. Are humans good intuitive statisticians after all? Rethinking some conclusions from the literature on judgment under uncertainty. *Cognition* 58: 1–73.

2005. Neurocognitive adaptations designed for social exchange. In *The Handbook of Evolutionary Psychology*, ed. D. Buss, 584–627. Hoboken, NJ: Wiley.

Coyne, J. A., N. H. Barton, and M. Turelli. 1997. Perspective: a critique of Sewall Wright's shifting balance theory of evolution. *Evolution* 51(3): 643–71.

Coyne, J. A., and H. A. Orr. 2004. *Speciation*. Sunderland, Mass.: Sinauer.

Cuvier, G. 1813. *Essay on the Theory of the Earth*, trans. Robert Kerr. Edinburgh: W. Blackwood.

Darwin, C. 1859. *On the Origin of Species by Means of Natural Selection, or the Preservation of Favoured Races in the Struggle for Life*. London: John Murray.

1861. *Origin of Species*, 3rd edn. London: John Murray.

1871. *The Descent of Man, and Selection in Relation to Sex*. London: John Murray.

1872. *On the Origin of Species*, 6th edn. London: John Murray.

1958. *The Autobiography of Charles Darwin 1809–1882. With the Original Omissions Restored. Edited and with Appendix and Notes by his Grand-Daughter Nora Barlow*. London: Collins.

1985–. *The Correspondence of Charles Darwin*. Cambridge University Press.

Darwin, E. 1794–96. *Zoonomia; or, The Laws of Organic Life*. London: J. Johnson.

1803. *The Temple of Nature*. London: J. Johnson.

Darwin, F. 1887. *The Life and Letters of Charles Darwin, Including an Autobiographical Chapter*. London: John Murray.

Davenport, C. B. 1930. The mingling of races. In *Human Biology and Racial Welfare*, ed. E. V. Cowdry. College Park, Md.: McGrath.

Davies, N. B. 1992. *Dunnock Behaviour and Social Evolution*. Oxford University Press.

Dawkins, R. 1976. *The Selfish Gene*. Oxford University Press.

1982. *The Extended Phenotype: The Gene as the Unit of Selection*. Oxford: W.H. Freeman.

1983. Universal Darwinism. In *Evolution from Molecules to Men*, ed. D. S. Bendall, 403–25. Cambridge University Press.

1986. *The Blind Watchmaker*. New York: Norton.

1989. The evolution of evolvability. In *Artificial Life*, ed. C. G. Langton, 201–20. Redwood City, Calif.: Addison-Wesley.

1992. Progress. In *Keywords in Evolutionary Biology*, ed. E. F. Keller, and E. Lloyd, 263–72. Cambridge, Mass.: Harvard University Press.

1996. *Climbing Mount Improbable*. New York: Norton.

1997. Human chauvinism: review of *Full House* by Stephen Jay Gould. *Evolution* 51(3): 1015–20.

2003. *A Devil's Chaplain: Reflections on Hope, Lies, Science and Love*. Boston: Houghton Mifflin.

2007. *The God Delusion*. New York: Houghton, Mifflin, Harcourt.

Dawkins, R., and J. R. Krebs. 1979. Arms races between and within species. *Proceedings of the Royal Society of London, Series B: Biological Sciences* 205: 489–511.

De Waal, F. 1982. *Chimpanzee Politics: Power and Sex among Apes*. London: Cape.

Dennett, D. C. 1990. Memes and the exploration of imagination. *Journal of Aesthetics and Art Criticism* 48: 127–35.

2006. *Breaking the Spell: Religion as a Natural Phenomenon*. New York: Viking.

Desmond, A. 1994. *Huxley, the Devil's Disciple*. London: Michael Joseph.

1997. *Huxley, Evolution's High Priest*. London: Michael Joseph.

Dick, S. J. 1996. *The Biological Universe: The Twentieth-Century Extraterrestrial Life Debate and the Limits of Science*. Cambridge University Press.

Diderot, D., 1943. *Diderot: Interpreter of Nature*. New York: International Publishers.

Dixon, T. 2008. *The Invention of Altruism: Making Moral Meanings in Victorian Britain*. Oxford University Press.

Dobzhansky, T. 1937. *Genetics and the Origin of Species*. New York: Columbia University Press.

1943. Temporal changes in the composition of populations of Drosophila pseudoobscura in different environments. *Genetics* 28: 162–86.

1962. *Mankind Evolving*. New Haven: Yale University Press.

1973. Nothing in biology makes sense except in the light of evolution. *American Biology Teacher* 35: 125–29.

Dörner, G. 1976. *Hormones and Brain Differentiation*. Amsterdam: Elsevier.

Dover, K. J. 1978. *Greek Homosexuality*. Cambridge, Mass.: Harvard University Press.

Downes, S. 2010. The basic components of the human mind were not solidified during the Pleistocene epoch. In *Contemporary Debates in the Philosophy of Biology*, ed. F. J. Ayala and R. Arp, 243–52. Malden, Mass.: Wiley-Blackwell.

Dudley, J. W. 1977. Seventy-six generations of selection for oil and protein percentages in maize. In *Proceedings of the International Conference on Quantitative*

Genetics, ed. E. Pollak, O. Kempthorne, and T. B. Bailey, 459–73. Ames: Iowa State University Press.

Edmonds, B. 2002. Three challenges for the survival of memetics. *Journal of Memetics: Evolutionary Models of Information Transmission* 6(2), http://cfpm.org/jom-emit/

Edwards, A. W. F. 2003. Human genetic diversity: Lewontin's fallacy. *BioEssays* 25: 798–801.

Eldredge, N., and S. J. Gould. 1972. Punctuated equilibria: an alternative to phyletic gradualism. In *Models in Paleobiology*, ed. T. J. M. Schopf, 82–115. San Francisco: Freeman, Cooper.

Endler, J. 1986. *Natural Selection in the Wild*. Princeton University Press.

Engelhardt, H. T. 1976. Ideology and etiology. *Journal of Medicine and Philosophy* 1: 256–68.

Erskine, F. 1995. "The Origin of Species" and the science of female inferiority. In *Charles Darwin's "The Origin of Species": New Interdisciplinary Essays*, ed. D. Amigoni and J. Wallace, 95–121. Manchester University Press.

Farlow, J. O., C. V. Thompson, and D. E. Rosner. 1976. Plates of the dinosaur Stegosaurus: forced convection heat loss fins? *Science* 192: 1123–25.

Fisher, R. A. 1930. *The Genetical Theory of Natural Selection*. Oxford University Press.

Fodor, J. 1983. *The Modularity of Mind*. Cambridge, Mass.: MIT Press.

1996. Peacocking. *London Review of Books* 18(8) April 18: 19–20.

2007. Why pigs don't have wings. The case against natural selection. *London Review of Books* 29(20) October 18: 19–22.

Fodor, J., and M. Piattelli-Palmarini. 2010. *What Darwin Got Wrong*. New York: Farrar, Straus, and Giroux.

Ford, E. B. 1964. *Ecological Genetics*. London: Methuen.

Foucault, M. 1978. *History of Sexuality, I*. New York: Pantheon.

Freud, S. [1905] 1955. *Three Essays on the Theory of Sexuality*. Standard Edition of the Complete Psychological Works of Sigmund Freud, vol. VII, 125–243. London: Hogarth.

Gadamer, H.-G. 1996. *The Enigma of Health*. Stanford University Press.

Galton, F. 1869. *Hereditary Genius*. London: Macmillan.

Gee, H. 1996. Box of bones "clinches" identity of Piltdown palaeontology hoaxer. *Nature* 382: 261–62.

Ghiselin, M. T. 1974. A radical solution to the species problem. *Systematic Zoology* 23: 536–44.

Gibbons, A. 2010. Tracing evolution's recent fingerprints. *Science* 329: 740–42.

Giere, R. 1988. *Explaining Science: A Cognitive Approach*. University of Chicago Press.

Gilbert, S. F., J. M. Opitz, and R. A. Raff. 1996. Resynthesizing evolutionary and developmental biology. *Developmental Biology* 173: 357–72.

Glasgow, J. M. 2003. On the new biology of race. *Journal of Philosophy* 100: 456–74.

Gluckman, P., A. Beedle, and M. Hanson. 2009. *Principles of Evolutionary Medicine*. Oxford University Press.

Goodman, N. 1955. *Fact, Fiction, and Forecast*. Cambridge, Mass.: Harvard University Press.

Gosse, P. 1857. *Omphalos; An Attempt to Untie the Geological Knot*. London: John Van Voorst.

Gould, S. J. 1981. *The Mismeasure of Man*. New York: Norton.

1988. On replacing the idea of progress with an operational notion of directionality. In *Evolutionary Progress*, ed. M. H. Nitecki, 319–38. University of Chicago Press.

1989. *Wonderful Life: The Burgess Shale and the Nature of History*. New York: Norton.

1996. *Full House: The Spread of Excellence from Plato to Darwin*. New York: Paragon.

2002. *The Structure of Evolutionary Theory*. Cambridge, Mass.: Harvard University Press.

Gould, S. J., and R. C. Lewontin. 1979. The spandrels of San Marco and the Panglossian paradigm: a critique of the adaptationist programme. *Proceedings of the Royal Society of London, Series B: Biological Sciences* 205: 581–98.

Grant, B. R., and P. R. Grant. 1989. *Evolutionary Dynamics of a Natural Population: The Large Cactus Finch of the Galapagos*. Chicago: University of Chicago Press.

Grant, M. 1916. *The Passing of the Great Race, or The Racial Basis of European History*. New York: Charles Scribner's Sons.

Grant, P. R. 1986. *Ecology and Evolution of Darwin's Finches*. Princeton University Press.

Grant, P. R., and B. R. Grant. 1995. Predicting microevolutionary responses to directional selection on heritable variation. *Evolution* 49: 241–51.

2007. *How and Why Species Multiply: The Radiation of Darwin's Finches*. Princeton University Press.

Gray, A. 1876. *Darwiniana*. New York: D. Appleton ; reprt., ed. A. H. Dupree, Cambridge, Mass. Harvard University Press, 1963.

Green, R. 1974. *Sexual Identity Conflict in Children and Adults*. New York: Basic Books.

Green, R. E., J. Kraus, S. E. Ptak, *et al.* 2006. Analysis of one million base pairs of Neanderthal DNA. *Nature* 444: 330–36.

Greene, J. C. 1959. *The Death of Adam: Evolution and Its Impact on Western Thought*. Ames: Iowa State University Press.

Greene, J. D., and J. Haidt. 2002. How (and where) does moral judgment work? *Trends in Cognitive Science* 6: 517–23.

Greene, J. D., R. B. Sommerville, L. E. Nystrom, J. M. Darley, and J. D. Cohen. 2001. An fRMI investigation of emotional engagement in moral judgment. *Science* 293: 2105–108.

Haeckel, E. 1896. *The Evolution of Man*. New York: Appleton.

Haig, D. 2008. Intimate relations: evolutionary conflicts of pregnancy and childhood. *Evolution in Health and Disease*, 2nd edn., ed. S. C. Sterns and J. C. Koella, 65–76. Oxford University Press.

Haldane, J. B. S. 1932. *The Causes of Evolution*. Ithaca, NY: Cornell University Press.

Hamer, D., and P. Copeland. 1994. *The Science of Desire: The Search for the Gay Gene and the Biology of Behavior*. New York: Simon & Schuster.

Hamer, D. H., S. Hu, V. L. Magnuson, Hu N., and A. M. L. Pattatucci. 1993. A linkage between DNA markers on the X-chromosome and male sexual orientation. *Science* 261: 321–37.

Hamilton, W. D. 1964. The genetical evolution of social behaviour. *Journal of Theoretical Biology* 7: 1–52.

   1967. Extraordinary sex ratios. *Science* 156: 477–88.

Hamilton, W. D., R. Axelrod, and R. Tanese. 1990. Sexual reproduction as an adaptation to resist parasites. *Proceedings of the National Academy of Sciences* 87(9): 3566–73.

Hamilton, W. D., and M. Zuk. 1982. Heritable true fitness and bright birds: a role for parasites. *Science* 218: 384–87.

Hardimon, M. O. 2003. The ordinary concept of race. *Journal of Philosophy* 100: 437–55.

Harvey, P., and R. May. 1989. Out for the sperm count. *Nature* 337: 508–509.

Hauser, M. D. 2006. The liver and the moral organ. *Social Cognitive and Affective Neuroscience* 1(3): 214–20.

Hempel, C. G. 1966. *Philosophy of Natural Science*. Englewood Cliffs, NJ: Prentice-Hall.

Hennig, W. 1966. *Phylogenetic Systematics*. Urbana, Ill.: University of Illinois Press.

Herre, E. A., C. A. Machado, and S. A. West. 2001. Selective regime and fig wasp sex ratios: toward sorting rigor from pseudo-rigor in tests of adaptation. In *Adaptation and Optimality*, ed. S. H. Orzack and E. Sober, 191–218. Cambridge University Press.

Hershberger, S. L. 2001. Biological factors in the development of sexual orientation. In *Lesbian, Gay, and Bisexual Identities and Youth: Psychological Perspectives*, ed. A. R. Patterson and C. J. D'Augelli, 27–51. Oxford University Press.

Hitchens, C. 2007. *God Is Not Great: How Religion Poisons Everything*. New York: Hachette.

Hitler, A. 1925. *Mein Kampf*, vol. I. Munich: Eher Verlag.

Holden, C. 2004. The origin of speech. *Science* 303: 1316–19.

Hopson, J. A. 1977. Relative brain size and behavior in archosaurian reptiles. *Annual Review of Ecology and Systematics* 8: 429–48.

Howells, W. W. 1960. The distribution of man. *Scientific American* 203(3): 112–27.

Hrdy, S. B. 1981. *The Woman That Never Evolved*. Cambridge, Mass.: Harvard University Press.

1999. *Mother Nature: A History of Mothers, Infants, and Natural Selection*. New York: Pantheon Books.

Hull, D. L., ed. 1973. *Darwin and His Critics*. Cambridge, Mass.: Harvard University Press.

1978. A matter of individuality. *Philosophy of Science* 45: 335–60.

1988. *Science as a Process*. University of Chicago Press.

Hume, D. [1739/40] 1940. *A Treatise of Human Nature*, ed. L. A. Selby-Bigge. Oxford University Press.

[1739/40] 1978. *A Treatise of Human Nature*, ed. P. H. Nidditch. Oxford University Press.

[1757] 1963. *The Natural History of Religion*. In *Hume on Religion*, ed. R. Wollheim. London: Fontana, 31–98.

Huxley, J. S. 1912. *The Individual in the Animal Kingdom*. Cambridge University Press.

1927. *Religion without Revelation*. London: Ernest Benn.

1942. *Evolution: The Modern Synthesis*. London: Allen and Unwin.

1943. *TVA: Adventure in Planning*. London: Scientific Book Club.

1959. Introduction, in Teilhard de Chardin, *The Phenomenon of Man*, 11–28. London: Collins.

Huxley, T. H. 1863. *Evidence as to Man's Place in Nature*. London: Williams and Norgate.

1870. On the geographical distribution of the chief modifications of mankind. *Journal of the Ethnological Society of London* 2: 404–12.

1874. On the hypothesis that animals are automata, and its history. In *Collected Essays*, vol. I, *Methods and Results*, 195–250. London: Macmillan.

[1893] 2009. *Evolution and Ethics*, ed. Michael Ruse. Princeton University Press.

Isaac, G. 1983. Aspects of human evolution. In *Evolution from Molecules to Men*, ed. D. S. Bendall, 509–43. Cambridge University Press.

James, W. 1880a. *The Principles of Psychology*. New York: Henry Holt.

1880b. Great men, great thoughts, and the environment. *Atlantic Monthly* 46(276): 441–59.

Jerison, H. 1973. *Evolution of the Brain and Intelligence*. New York: Academic Press.

Johanson, D., and M. Edey. 1981. *Lucy: The Beginnings of Humankind*. New York: Simon & Schuster.

Johnson, P. E. 1991. *Darwin on Trial*. Washington, D.C.: Regnery Gateway.

Jordan-Young, R. M. 2010. *Brainstorm: The Flaws in the Science of Sex Differences*. Cambridge, Mass.: Harvard University Press.

Kant, I. [1785] 1959. *Foundations of the Metaphysics of Morals*, trans. Lewis White Beck. Indianapolis: Bobbs-Merrill.

[1790] 1928. *The Critique of Teleological Judgement*, trans. J. C. Meredith. Oxford University Press.

Kershaw, I. 1999. *Hitler 1889–1936: Hubris*. New York: Norton.

Kevles, D. J. 1985. *In the Name of Eugenics: Genetics and the Uses of Human Heredity*. New York: Knopf.

Kimura, M. 1983. *The Neutral Theory of Molecular Evolution*. Cambridge University Press.

Kinsey, A. C., W. B. Pomeroy, and C. E. Martin. 1948. *Sexual Behavior in the Human Male*. Philadelphia: W. B. Saunders.

Kinsey, A. C., W. B. Pomeroy, C. E. Martin, and P. H. Gebhard. 1953. *Sexual Behavior in the Human Female*. Philadelphia: W. B. Saunders.

Kitcher, P. 1997. *The Lives to Come: The Genetic Revolution and Human Possibilities*, 2nd edn. New York: Simon & Schuster.

1999. Race, ethnicity, biology, culture. In *In Mendel's Mirror*, 230–57. New York: Oxford University Press.

2007. Does "race" have a future? *Philosophy and Public Affairs* 35: 293–317.

Knoll, A. 2003. *Life on a Young Planet: The First Three Billion Years of Evolution on Earth*. Princeton University Press.

Krings, M., A. Stone, R. W. Schmitz, H. Krainitzki, M. Stoneking, and S. Pääbo. 1997. Neanderthal DNA sequences and the origin of modern humans. *Cell* 90: 19–30.

Kropotkin, P. 1902. *Mutual Aid: A Factor in Evolution*. Boston: Extending Horizons Books.

Kuhn, T. 1962. *The Structure of Scientific Revolutions*. University of Chicago Press.

   1969. *The Structure of Scientific Revolutions*, 2nd edn. University of Chicago Press.

Kuper, A. 2000. If memes are the answer, what is the question? In *Darwinizing Culture*, ed. R. Aunger, 175–88. Oxford University Press.

Larsen, C. S. 2008. *Our Origins: Discovering Physical Anthropology*. New York: Norton.

Larson, E. J. 1997. *Summer for the Gods: The Scopes Trial and America's Continuing Debate over Science and Religion*. New York: Basic Books.

Laumann,, E. O., Gagnon,, J. H., Michael,, R. T., Michaels,, S.1994. *The Social Organization of Sexuality: Sexual Practices in the United States*. University of Chicago Press.

Laxova, R. 1998. Lionel Sharples Penrose, 1898–1972: a personal memoir in celebration of the centenary of his birth. *Genetics* 150: 1333–40.

LeVay, S. 2010. *Gay, Straight, and the Reason Why: The Science of Sexual Orientation*. Oxford University Press.

Levins, R., and R. C. Lewontin. 1985. *The Dialectical Biologist*. Cambridge, Mass.: Harvard University Press.

Lewens, T. 2007. *Darwin*. London: Routledge.

Lewin, R. 1987. *Bones of Contention: Controversies in the Search for Human Origins*. New York: Simon & Schuster.

   1989. *Human Evolution: An Illustrated Introduction*, 2nd edn. Oxford: Blackwell Scientific.

Lewontin, R. C. 1982. *Human Diversity*. New York: Scientific American Library.

   2000. *The Triple Helix: Gene, Organism, and Environment*. Cambridge, Mass.: Harvard University Press.

Lieberman, P. 1984. *The Biology and Evolution of Language*. Cambridge, Mass.: Harvard University Press.

Lipton, P. 1991. *Inference to the Best Explanation*. London: Routledge.

Livingstone, D. N. 2008. *Adam's Ancestors: Race, Religion, and the Politics of Human Origins*. Baltimore: Johns Hopkins University Press.

London, J. 1903. *The Call of the Wild*. New York: Macmillan.

Lorenz, K. 1941. Kant's Lehre vom Apriorischen im Lichte gegenwärtiger Biologie. *Blätter für deutsche Philosophie* 15: 94–125. Trans. and reprinted as: Kant's doctrine of the "a priori" in the light of contemporary biology. *General Systems* 7 (1962), 23–35. Reprinted in Ruse 2009b, 231–42 (to which page numbers in the text refer).

Lovejoy, A. O. 1936. *The Great Chain of Being*. Cambridge, Mass.: Harvard University Press.

Lovejoy, C. O. 1981. The origin of man. *Science* 211: 341–50.

Mackie, J. 1966. The direction of causation. *Philosophical Review* 75: 441–66.

 1977. *Ethics*. Harmondsworth: Penguin Books.

 1979. *Hume's Moral Theory*. London: Routledge and Kegan Paul.

Majerus, M. E. N. 1998. *Melanism: Evolution in Action*. Oxford University Press.

Martin, L. G. 1999. What is the function of the human appendix? Did it once have a purpose that has since been lost?. *Scientific American*, October 21.

Matthew, W. D. 1915. Climate and evolution. *Annals of the New York Academy of Sciences* 24: 171–318.

 1939. *Climate and Evolution*. New York Academy of Sciences.

Maynard Smith, J. 1978. *The Evolution of Sex*. Cambridge University Press.

 1981. Did Darwin get it right? *London Review of Books* 3(11) June 18: 10–11.

 1982. *Evolution and the Theory of Games*. Cambridge University Press.

Maynard Smith, J., and E. Szathmáry. 1995. *The Major Transitions in Evolution*. New York: Oxford University Press.

Mayr, E. 1942. *Systematics and the Origin of Species*. New York: Columbia University Press.

 1963. *Animal Species and Evolution*. Cambridge, Mass.: Harvard University Press.

 1988. *Toward a New Philosophy of Biology: Observations of an Evolutionist*. Cambridge, Mass.: Belknap Press of Harvard University Press.

Medawar, P. 1967. Review of *The Phenomenon of Man*. In *The Art of the Soluble*, ed. P. Medawar. London: Methuen. Originally published in *Mind* 70 (1961): 99–106.

McDonald, J. H., G. K. Chambers, J. David, and F. J. Ayala. 1977. Adaptive response due to changes in gene regulation: a study with *Drosophila*. *Proceedings of the National Academy of Sciences* 74: 4562–66.

McGinn, C. 2000. *The Mysterious Flame: Conscious Minds in a Material World*. New York: Basic Books.

McHenry, H. M. 2009. Human evolution. In *Evolution: The First Four Billion Years*, ed. M. Ruse and J. Travis, 256–80. Cambridge, Mass.: Harvard University Press.

McShea, D. W. 1991. Complexity and evolution: what everybody knows. *Biology and Philosophy* 6(3): 303–25.

McShea, D. W., and R. N. Brandon. 2010. *Biology's First Law: The Tendency for Diversity and Complexity to Increase in Evolutionary Systems.* University of Chicago Press.

Mill, J. S. 1840. Review of the works of Samuel Taylor Coleridge. *London and Westminster Review* 33: 257–302.

　　1863. *Utilitarianism.* London: Parker, Son, and Bourn.

Miller, E. M. 2000. Homosexuality, birth order, and evolution: toward an equilibrium reproductive economics of homosexuality. *Archives of Sexual Behavior* 29: 1–34.

Mithen, S. 1996. *The Prehistory of the Mind.* London: Thames and Hudson.

Money, J., and A. Ehrhardt. 1972. *Man and Woman, Boy and Girl: The Differentiation and Dimorphism of Gender Identity from Conception to Maturity.* Baltimore: Johns Hopkins University Press.

Montagu, A. 1942. *Man's Most Dangerous Myth: The Fallacy of Race.* New York: Columbia University Press.

Montagu, A., E. Beaglehole, J. Comas, *et al.* 1950. *The Race Question. Statement Issued July 18, 1950.* Paris: Unesco.

Moore, G. E. 1903. *Principia Ethica.* Cambridge University Press.

Murphy, D. 2008. Concepts of health and disease. In *The Stanford Encyclopedia of Philosophy,* summer 2009 edn., ed. Edward N. Zalta, http://plato.stanford.edu//entries/health-disease/

Murphy, J. 1982. *Evolution, Morality, and the Meaning of Life.* Totowa, NJ: Rowman and Littlefield.

Nagel, E. 1961. *The Structure of Science: Problems in the Logic of Scientific Explanation.* New York: Harcourt, Brace, and World.

Nesse, R. M., and G. C. Williams. 1994. *Why We Get Sick: The New Science of Darwinian Medicine.* New York: Times Books.

　　1995. *Evolution and Healing: The New Science of Darwinian Medicine.* London: Weidenfeld and Nicholson.

Nietzsche, F. 1882. *Vorspiel einer Philosophie der Zukunft*; trans. Walter Kaufmann as *The Gay Science.* New York: Random House, 1974.

　　1886. *Jenseits von Gut und Böse: Vorspiel einer Philosophie der Zukunft* [Beyond Good and Evil: Prelude to a Philosophy of the Future]. Leipzig: Neumann.

　　1887. *Zur Genealogie der Moral: Eine Streitschrift* [On the Genealogy of Morality: A Polemic]. Leipzig: Neumann.

Oster, G., and E. O. Wilson. 1978. *Caste and Ecology in the Social Insects.* Princeton University Press.

Owen, R. 1848. *On the Archetype and Homologies of the Vertebrate Skeleton.* London: John Van Voorst.

1849. *On the Nature of Limbs*. London: John Van Voorst.

1861. *Paleontology*, 2nd edn. Edinburgh: Black.

Parker, G. A. 1978. Evolution of competitive mate searching. *Annual Review of Entomology* 23: 173–96.

Pasterniani, E. 1969. Selection for reproductive isolation between two populations of maize, *Zea mays L. Evolution* 23: 534–47.

Peirce, C. S. [1892] 1958. Conclusion of the history of science lectures [Lowell Lectures on the History of Science]. In *Values in a World of Chance: Selected Writings of Charles S. Peirce (1839–1914*, ed. P. P. Wiener, 257–60. Garden City, NY: Doubleday.

1997. *Pragmatism as a Principle and Method of Right Thinking: The 1903 Harvard Lectures on Pragmatism*, ed. P. A. Turrisi. Albany, NY: State University of New York Press.

Pigliucci, M., and J. Kaplan. 2003. On the concept of biological race and its applicability to humans. *Philosophy of Science* 70: 1161–72.

Pinker, S. 1994. *The Language Instinct: How the Mind Creates Language*. New York: William Morrow.

1997. *How the Mind Works*. New York: Norton.

Plantinga, A. 1993. *Warrant and Proper Function*. New York: Oxford University Press.

2009. Science and religion: why does the debate continue? In *The Religion and Science Debate: Why Does It Continue?*, ed. H. W. Attridge, 93–123. New Haven: Yale University Press.

Popper, K. R. 1963. *Conjectures and Refutations*. London: Routledge and Kegan Paul.

1972. *Objective Knowledge*. Oxford University Press.

1974. Intellectual autobiography. In *The Philosophy of Karl Popper*, ed. Paul A. Schilpp, vol. I, 3–181. LaSalle, Ill.: Open Court.

Provine, W. B. 1971. *The Origins of Theoretical Population Genetics*. University of Chicago Press.

1986. *Sewall Wright and Evolutionary Biology*. University of Chicago Press.

Putnam, H. 1981. *Reason, Truth, and History*. Cambridge University Press.

Pyysiäinen, I., and M. Hauser. 2010. The origins of religion: evolved adaptation or by-product? *Trends in Cognitive Sciences* 14(3): 104–109.

Quine, W. V. O. 1969. *Ontological Relativity and Other Essays*. New York: Columbia University Press.

Raup, D. 1988. Testing the fossil record for evolutionary progress. In *Evolutionary Progress*, ed. M. Nitecki, 293–317. University of Chicago Press.

Rawls, J. 1971. *A Theory of Justice*. Cambridge, Mass. Harvard University Press.

Reich, D., R. E. Green, M. Kircher, *et al.* 2010. Genetic history of an archaic hominin group from Denisova Cave in Siberia. *Nature* 468: 1053–60.

Relethford, J. H. 2008. *The Human Species: An Introduction to Biological Anthropology*, 7th edn. Boston: McGraw-Hill.

Reynolds, V., and R. Tanner. 1983. *The Biology of Religion*. London: Longman.

Reznek, L. 1987. *The Nature of Disease*. London: Routledge.

Richards, R. J. 1987. *Darwin and the Emergence of Evolutionary Theories of Mind and Behavior*. University of Chicago Press.

   1992. *The Meaning of Evolution: The Morphological Construction and Ideological Reconstruction of Darwin's Theory*. University of Chicago Press.

   2003. *The Romantic Conception of Life: Science and Philosophy in the Age of Goethe*. University of Chicago Press.

   2008. *The Tragic Sense of Life: Ernst Haeckel and the Struggle over Evolutionary Thought*. University of Chicago Press.

Richerson, P., and R. Boyd. 2005. *Not by Genes Alone: How Culture Transformed Human Evolution*. University of Chicago Press.

Romanes, E., ed. 1895. *The Life and Letters of George John Romanes*. London: Longmans, Green.

Rorty, R. 1979. *Philosophy and the Mirror of Nature*. Princeton University Press.

Rosenberg, A., and D. McShea. 2008. *Philosophy of Biology: A Contemporary Discussion*. London: Routledge.

Rosenberg, N. A., J. K. Pritchard, J. L. Weber, *et al.* 2002. Genetic structure of human populations. *Science* 298: 2381–85.

Roughgarden, J. 2009. *Evolution's Rainbow: Diversity, Gender, and Sexuality in Nature and People*. Berkeley: University of California Press.

Roughgarden, J., M. Oishi, and E. Akçay. 2006. Reproductive social behavior: cooperative games to replace sexual selection. *Science* 311: 965–69.

Rupke, N. A. 1994. *Richard Owen: Victorian Naturalist*. New Haven: Yale University Press.

Ruse, M. 1973. *The Philosophy of Biology*. London: Hutchinson.

   1979. *Sociobiology: Sense or Nonsense?* Dordrecht: Reidel.

   1980. Charles Darwin and group selection. *Annals of Science* 37: 615–30.

   1986. *Taking Darwin Seriously: A Naturalistic Approach to Philosophy*. Oxford: Blackwell.

   1987a. Biological species: natural kinds, individuals, or what? *British Journal for the Philosophy of Science* 38: 225–42.

   1987b. Darwin and determinism. *Zygon* 22: 419–42.

Ruse, M. ed. 1988a. *But Is It Science? The Philosophical Question in the Creation/ Evolution Controversy*. Buffalo, NY: Prometheus.

Ruse, M. 1988b. *Homosexuality: A Philosophical Inquiry*. Oxford: Blackwell.

1991. William Whewell: omniscientist. In *William Whewell: A Composite Portrait*, ed. M. Fisch, and S. Shaffer, 87–116. Oxford University Press.

1996. *Monad to Man: The Concept of Progress in Evolutionary Biology.* Cambridge, Mass.: Harvard University Press.

1999. *Mystery of Mysteries: Is Evolution a Social Construction?* Cambridge, Mass.: Harvard University Press.

2001. *Can a Darwinian Be a Christian? The Relationship between Science and Religion.* Cambridge University Press.

2003. *Darwin and Design: Does Evolution Have a Purpose?* Cambridge, Mass.: Harvard University Press.

2004. Adaptive landscapes and dynamic equilibrium: the Spencerian contribution to twentieth-century American evolutionary biology. In *Darwinian Heresies*, ed. A. Lustig, R. J. Richards, and M. Ruse, 131–50. Cambridge University Press.

2005a. Darwin and mechanism: metaphor in science. *Studies in History and Philosophy of Biology and Biomedical Sciences* 36: 285–302.

2005b. *The Evolution–Creation Struggle.* Cambridge, Mass.: Harvard University Press.

2008. *Charles Darwin.* Oxford: Blackwell.

2009a. The Darwinian revolution: rethinking its meaning and significance. *Proceedings of the National Academy of Sciences* 106: 10040–47.

Ruse, M.ed. 2009b. *Philosophy after Darwin: Classic and Contemporary Readings.* Princeton University Press.

Ruse, M. 2010. *Science and Spirituality: Making Room for Faith in the Age of Science.* Cambridge University Press.

Ruse, M., and E. O. Wilson. 1986. Moral philosophy as applied science. *Philosophy* 61: 173–92.

Russell, B. [1937] 2004. *Power: A New Social Analysis.* London: Routledge.

Russell, R. J. 2008. *Cosmology: From Alpha to Omega. The Creative Mutual Interaction of Theology and Science.* Minneapolis: Fortress Press.

Schaffner, K. 1993. *Discovery and Explanation in Biology and Medicine.* University of Chicago Press.

Sedley, D. 2008. *Creationism and Its Critics in Antiquity.* Berkeley: University of California Press.

Sepkoski, D., and M. Ruse, eds. 2009. *The Paleobiological Revolution.* University of Chicago Press.

Shipman, P. 2002. *The Man Who Found the Missing Link: Eugene Dubois and His Lifelong Quest to Prove Darwin Right.* Cambridge, Mass.: Harvard University Press.

Sidgwick, H. 1876. The theory of evolution in its application to practice. *Mind* 1: 52–67.

Simpson, G. G. 1944. *Tempo and Mode in Evolution*. New York: Columbia University Press.

1949. *The Meaning of Evolution*. New Haven: Yale University Press.

Singer, P. 2005. Ethics and intuitions. *Journal of Ethics* 9: 331–52.

Singh, D. 1993. Adaptive significance of female physical attractiveness: role of waist-to-hip ratio. *Journal of Personality and Social Psychology* 65: 293–307.

Sober, E. 1981. The evolution of rationality. *Synthese* 46: 95–120.

Spencer, F. 1990. *Piltdown: A Scientific Forgery*. New York: Oxford University Press.

Spencer, H. 1851. *Social Statics; or the Conditions Essential to Human Happiness Specified and the First of Them Developed*. London: J. Chapman.

1855. *Principles of Psychology*. London: Longman, Brown, Green, and Longmans.

1857. Progress: its law and cause. *Westminster Review* 67: 244–67.

1862. *First Principles*. London: Williams and Norgate.

Sperber, D. 2000. An objection to the memetic approach to culture. In *Darwinizing Culture*, ed. R. Aunger, 163–73. Oxford University Press.

Stebbins, G. L. 1950. *Variation and Evolution in Plants*. New York: Columbia University Press.

Sterelny, K. 2003. *Thought in a Hostile World: The Evolution of Human Cognition*. Oxford: Blackwell.

Stringer, C. 2002. Modern human origins – progress and prospects. *Philosophical Transactions of the Royal Society B* 357: 563–79.

2003. Human evolution: out of Ethiopia. *Nature* 423: 692–95.

Stringer, C., and P. Andrews. 2005. *The Complete World of Human Evolution*. New York: Thames and Hudson.

Sulloway, F. 1979. *Freud: Biologist of the Mind*. New York: Basic Books.

Sumner, W. G. 1914. *The Challenge of Facts and Other Essays*. New Haven: Yale University Press.

Sussman, R. W. 1991. Primate origins and the evolution of angiosperms. *American Journal of Primatology* 23: 209–23.

Swaab, D. F. 2008. Sexual orientation and its basis in brain structure and function. *Proceedings of the National Academy of Sciences* 105(30): 10273–74.

Szalay, F. S., and M. Dagosto. 1988. Evolution of hallucial grasping in the primates. *Journal of Human Evolution* 17: 1–33.

Teilhard de Chardin, P. 1955. *Le phénomène humain*. Paris: Éditions du Seuil.

Thompson, R. P. 1989. *The Structure of Biological Theories*. Albany, NY: State University of New York University Press.

Toulmin, S. 1967. The evolutionary development of science. *American Scientist* 57: 456–71.

Trivers, R. L. 1971. The evolution of reciprocal altruism. *Quarterly Review of Biology* 46: 35–57.

1974. Parent–offspring conflict. *American Zoologist* 14: 249–64.

2002. *Natural Selection and Social Theory: Selected Papers of Robert L. Trivers.* Oxford University Press.

Trivers, R. L., and D. E. Willard. 1973. Natural selection of parental ability to vary the sex ratio of offspring. *Science* 179: 90–92.

Tuana, N. 1993. *The Less Noble Sex: Scientific, Religious, and Philosophical Conceptions of Woman's Nature.* Bloomington: Indiana University Press.

Van den Berghe, P. 1979. *Human Family Systems.* New York: Elsevier.

Vermeij, G. J. 1987. *Evolution and Escalation.* Princeton University Press.

Vogel, S. 1988. *Life's Devices: The Physical World of Animals and Plants.* Princeton University Press.

von Bernhardi, F. 1912. *Germany and the Next War.* London: Edward Arnold.

Vorzimmer, P. J. 1970. *Charles Darwin: The Years of Controversy.* Philadelphia, Penn.: Temple University Press.

Wallace, A. R. 1870. *Contributions to the Theory of Natural Selection: A Series of Essays.* London: Macmillan.

1900. *Studies: Scientific and Social.* London: Macmillan.

1905. *My Life: A Record of Events and Opinions.* London: Chapman and Hall.

Weikart, R. 2004. *From Darwin to Hitler: Evolutionary Ethics, Eugenics, and Racism in Germany.* New York: Palgrave Macmillan.

Weinberg, S. 1977. *The First Three Minutes: A Modern View of the Origin of the Universe.* New York: Basic Books.

Westermarck, E. 1906. *The Origin and Development of the Moral Ideas,* 2 vols. London: Macmillan.

Wheeler, P. E. 1984. The evolution of bipedality and loss of functional body hair in hominoids. *Journal of Human Evolution* 13: 91–98.

Whewell, W. 1840. *The Philosophy of the Inductive Sciences.* London: Parker.

2001. *Of the Plurality of Worlds. A Facsimile of the First Edition of 1853: Plus Previously Unpublished Material Excised by the Author Just Before the Book Went to Press; and Whewell's Dialogue Rebutting His Critics, Reprinted from the Second Edition,* ed. M. Ruse. University of Chicago Press.

White, T. D., B. Asfaw, Y. Beyene, *et al.* 2009. *Ardipithecus ramidus* and the paleobiology of early hominids. *Science* 326(5949): 75–86.

Williams, G. C. 1966. *Adaptation and Natural Selection.* Princeton University Press.

1975. *Sex and Evolution*. Princeton University Press.

Wilson, D. S. 2002. *Darwin's Cathedral: Evolution, Religion, and the Nature of Society*. University of Chicago Press.

Wilson, E. O. 1975. *Sociobiology: The New Synthesis*. Cambridge, Mass.: Harvard University Press.

1978. *On Human Nature*. Cambridge, Mass. Harvard University Press.

1980a. Caste and division of labor in leaf cutter ants (Hymenoptera: Formicidae: *Atta*). I The overall pattern in *Atta sexdens*. *Behavioral Ecology and Sociobiology* 7: 143–56.

1980b. Caste and division of labor in leaf cutter ants (Hymenopter: Formicidae: *Atta*). II The ergonomic optimization of leaf cutting. *Behavioral Ecology and Sociobiology* 7: 157–65.

1984. *Biophilia*. Cambridge, Mass.: Harvard University Press.

1992. *The Diversity of Life*. Cambridge, Mass.: Harvard University Press.

2002. *The Future of Life*. New York: Vintage Books.

2006. *The Creation: A Meeting of Science and Religion*. New York: Norton.

Wilson, G. D., and Q. Rahman. 2005. *Born Gay: The Psychobiology of Sex Orientation*. London: Peter Owen.

Wittgenstein, L. 1923. *Tractatus Logico-Philosophicus*. London: Routledge and Kegan Paul.

Woese, C. R. 1998. The universal ancestor. *Proceedings of the National Academy of Sciences* 95: 6854–59.

Wolpoff, M., and R. Caspari. 1997. *Race and Human Evolution*. Boulder, Colo.: Westview.

World Health Organization (WHO). 1946. WHO definition of health. In *Preamble to the Constitution of the World Health Organization as Adopted by the International Health Conference*. Geneva: WHO.

Wright, C. 1877. The evolution of self consciousness. In *Philosophical Discussions*, 199–266. New York: Henry Holt.

Wright, S. 1931. Evolution in Mendelian populations. *Genetics* 16: 97–159.

1932. The roles of mutation, inbreeding, crossbreeding and selection in evolution. *Proceedings of the Sixth International Congress of Genetics* 1: 356–66.

Index

Made in the USA
Lexington, KY
15 January 2015